D1672067

UB MAGDEBURG
MA9
002 261 464

Non-Regular Differential Equations and Calculations of Electromagnetic Fields

Non-Regular Differential Equations and Calculations of Electromagnetic Fields

N E Tovmasyan
State Engineering University of Armenia
Armenia

edited by

L Z Gevorkyan
State Engineering University of Armenia
Armenia

M S Ginovyan
Institute of Mathematics
Armenian National Academy of Sciences
Armenia

M N Bobrova
State Engineering University of Armenia
Armenia

Published by

World Scientific Publishing Co. Pte. Ltd.

P O Box 128, Farrer Road, Singapore 912805

USA office: Suite 1B, 1060 Main Street, River Edge, NJ 07661

UK office: 57 Shelton Street, Covent Garden, London WC2H 9HE

Library of Congress Cataloging-in-Publication Data

Tovmasyan, N.E., 1934–
Non-regular differential equations and calculations of electromagnetic fields / N.E. Tovmasyan; edited by L.Z. Gevorkyan, M.S. Ginovyan, M.N. Bobrova.
p. cm.
Includes bibliographical references and index.
ISBN 9810233361 (alk. paper)
1. Electrodynamics -- Mathematics. 2. Differential equations, Partial -- Numerical solutions. 3. Boundary value problems -- Numerical solutions. I. Gevorkyan, L. Z. II. Ginovyan, M. S. III. Bobrova, M. N.
QC631.T684 1998
537.6'01'515353--dc21 98-10860
CIP

British Library Cataloguing-in-Publication Data

A catalogue record for this book is available from the British Library.

This book is printed on acid-free paper.

Printed in Singapore by Uto-Print

Preface

This monograph is devoted to the theory of partial differential equations in the plane domains and their applications in Electrodynamics.

The theory of partial differential equations in the plane domains possesses a number of essential peculiarities. The regular and non-regular, properly and improperly elliptic equations should be distinguished depending on the distribution of the roots of the corresponding characteristic equations. For non-regular and improperly elliptic equations the classical boundary value problems (Cauchy, Dirichlet, Neumann and Poincaré) are not correct.

In this monograph for non-regular differential equations and for improperly elliptic equations we pose and resolve correct boundary problems. The methods developed here are used for investigation of propagation of periodic electromagnetic waves in stratified and non-homogeneous media as well as for calculations of capacitances of a sufficiently wide class of new capacitors.

The book consists of 7 chapters.

In the first three chapters we consider boundary value problems for non-regular partial differential equations in the half-plane. Based on the non-regularity character of these equations we describe and resolve Riemann–Hilbert and Dirichlet-type problems. We describe some efficient methods of resolution of such problems for non-regular differential equations $L_1 L_2 \cdots L_n u = 0$, where $L_1, L_2, \ldots, L_n$ are first-order linear differential operators with constant coefficients.

In chapter 4 we investigate the propagation laws of periodic electromagnetic waves in the stratified and non-homogeneous media. We show that the electric and magnetic intensities can be explicitly expressed by the boundary data.

Chapter 5 is devoted to the calculations of capacitances of cylindrical and spherical capacitors. Using conformal mappings of plane domains we calculate the capacitances of a sufficiently broad class of cylindrical capacitors. Some problems of the best choice of the shape of these capacitors are described and resolved. For spherical capacitors we derive a new simple formula of capacitance.

In chapter 6, numerous boundary value problems for improperly elliptic equations in bounded plane domains are investigated. The assumption of impropriety of the equations has essential influence on the description and methods of solution of the problems. For these equations the Riemann–Hilbert and Dirichlet-type problems are of special significance. In the canonical domains (circle, ellipse, etc.), explicit formulae for solutions of these equations are obtained.

In chapter 7 we construct the fundamental systems of solutions for improperly elliptic equations in the class of real-valued functions. The results of this chapter are used in chapter 6 for investigation of boundary value problems for improperly elliptic equations.

Chapters 1, 2, 3 and 7 include the theoretical part of the book, while the chapters 4, 5 and 6 are devoted to the application of the results of theoretical part to the efficient resolutions of concrete applied problems.

The main advantages of this book are:

1. We have constructed classes of functions for which the Riemann-Hilbert and Dirichlet type boundary value problems for non-regular partial differential equations in the half-plane are correct (chapters 1, 2 and 3).

2. We suggest efficient methods of solution of a broad class of boundary value problems for improperly elliptic equations in finite simply connected and canonical domains (chapter 6).

3. The obtained results are applied for investigation of the propagation of periodic electromagnetic waves in stratified and non-homogeneous media, as well as for calculations of capacitances of capacitors with analytic surfaces (chapters 4 and 5).

Methods and results of this book may be applied for resolution of a broad class of important practical problems of mathematical physics that arise in the different fields of science and engineering.

The book is based mostly on the investigations of the author and the larger part of the results is published here for the first time.

The manuscript was read by scientific collaborators of the Department of Mathematics of Armenian State Engineering University and Institute of Mathematics of Armenian National Academy of Sciences. L. Z. Gevorkyan, M. S. Ginovyan, A. O. Babayan and A. A. Andryan have made a great number of useful remarks. Essential assistance in the preparation of the layout of this book was brought by M. N. Bobrova, A. P. Antonyan, A. N. Tovmasyan and S. M. Carapetyan. I would like to thank all of them.

I would like to express a special gratitude to His Holiness Karekin I, Catholicos of All Armenians for the financial support during the preparation of the manuscript.

N. E. Tovmasyan

Contents

Chapter 1

Dirichlet Type Problems for Non-regular Differential Equations in the Half-plane

1.1 Description of the Problem and Main Results

Consider the following differential equation

$$\frac{\partial^n u}{\partial t^n} + \sum_{k=0}^{n-1} \frac{\partial^k}{\partial t^k} P_k(i\frac{\partial}{\partial x})\, u = 0, \quad t > 0, \quad x \in R^1, \tag{1.1}$$

where $P_k(\sigma)$ are polynomials with constant coefficients on real variable σ, i is the imaginary unit, $u(x,t)$ is a solution to be found and R^1 is the real axis.

If $P(\sigma) = a_0 + a_1\sigma + \cdots + a_n\sigma^n$, then

$$P(i\frac{\partial}{\partial x})u \equiv a_0 u + a_1 i\frac{\partial u}{\partial x} + \cdots + a_n i^n \frac{\partial^n u}{\partial x^n}.$$

The equation

$$\lambda^n + \sum_{k=0}^{n-1} P_k(\sigma)\lambda^k = 0, \quad \sigma \in R^1 \tag{1.2}$$

with respect to the complex variable λ is called characteristic equation, corresponding to equation (1.1).

Below we will see that the correctness of the boundary value problems for equation (1.1) essentially depends on the number of the roots of characteristic equation (1.2) satisfying $\mathrm{Re}\lambda \leq 0$.

Let $\rho(\sigma)$ be the number of the roots of the characteristic equation (1.2) with $\mathrm{Re}\lambda \leq 0$ at the point $\sigma \in R^1$. The examples show that $\rho(\sigma)$, in general, depends on σ. For example, if

$$\frac{\partial u}{\partial t} + i\frac{\partial u}{\partial x} = 0, \tag{1.3}$$

then $\rho(\sigma)$ is determined by the formula

$$\rho(\sigma) = 1 \quad \text{for} \quad \sigma \geq 0 \quad \text{and} \quad \rho(\sigma) = 0 \quad \text{for} \quad \sigma < 0. \tag{1.4}$$

Equation (1.1) is said to be regular, if

$$\rho(\sigma) = r, \quad \sigma \in R^1, \quad \sigma \neq \sigma_1, \ldots, \sigma_m. \tag{1.5}$$

Here r is an integer, independent of σ $(0 \leq r \leq n)$ and $\sigma_1, \cdots, \sigma_m$ are some fixed points of R^1. These points may be absent and in this case $\rho(\sigma) = r$, $\sigma \in R^1$. The number r is called the order of regularity of equation (1.1).

For regular equation (1.1) the Cauchy problem in different classes of functions and distributions has been considered in [1] - [10]. Cauchy boundary data in this case have the form

$$\frac{\partial^k u(x,0)}{\partial t^k} = f_k(x), \quad x \in R^1, \quad k = 0, \ldots, r-1, \tag{1.6}$$

where $f_k(x)$ are known functions increasing not rapidly than polynomial at $|x| \to +\infty$ and r is the order of regularity of equation (1.1). A solution of the problem (1.1) is searched for in a similar class.

In [1] - [10] it has been shown that for regular equation (1.1) with order of regularity r the problem (1.1), (1.5) is solvable.

In the case where $\rho(\sigma) = r$, $\sigma \in R^1$ in the above mentioned papers it has been shown that the Cauchy problem (1.1), (1.6) is uniquely solvable, that is, a solution of the problem (1.1), (1.6) exists and is unique.

In this chapter we consider mainly boundary value problems for non-regular differential equations. We suppose that

$$\rho(\sigma) = r \quad \text{for} \quad \sigma > 0, \tag{1.7}$$

$$\rho(\sigma) = q \quad \text{for} \quad \sigma < 0, \tag{1.8}$$

where r and q are some integers, $0 \leq r \leq n, 0 \leq q \leq n$. For definiteness, we assume that $r \leq q$. In particular, equation (1.1) is regular if $r = q$.

As boundary conditions for equation (1.1) will be taken

$$\frac{\partial^k u(x,0)}{\partial t^k} = f_k(x), \quad x \in R^1, \quad k = 0, \dots, r-1, \tag{1.9}$$

$$Re\left[b_k(x)\frac{\partial^k u(x,0)}{\partial t^k}\right] = f_k(x), \quad x \in R^1, \quad k = r, \dots, q-1, \tag{1.10}$$

where $b_k(x)$ and $f_k(x)$ are given functions defined on R^1 and

$$b_k(x) \neq 0, \quad x \in R^1,\ k = r, \dots, q-1. \tag{1.11}$$

The problem (1.1), (1.9), (1.10) will be called homogeneous if $f_k(x) \equiv 0$ ($k = 0, \dots, q-1$). If $q = r$, conditions (1.10) are absent and if $r = 0$, conditions (1.9) are absent.

Without loss of generality we will assume that $|\, b_k(x) \,| = 1$ for $x \in R^1$. This may be realized by dividing both sides of (1.10) by $|\, b_k(x) \,|$.
Let R^2_+ be the half-plane $x \in R^1$, $t > 0$ and α be a non-negative constant.
Let us introduce the following notation: $M_\alpha(R^2_+)$ is the class of infinitely differentiable in R^2_+ functions $u(x,t)$, which for $k \geq 0$ and $j \geq 0$ satisfy

$$\left|\frac{\partial^{j+k} u(x,t)}{\partial x^j \partial t^k}\right| \leq c_{jk}(1+|\, x \,|)^{\alpha_{jk}}(1+t)^{\beta_{jk}}, \quad (x,t) \in R^2{}_+, \tag{1.12}$$

where c_{kj}, α_{kj}, and β_{kj} are some constants depending on $u(x,t)$ and $\alpha_{jk} < \alpha$.

Denote by $N_\alpha(R^2_+)$ the class of infinitely differentiable in R^2_+ functions $u(x,t)$ satisfying

$$\left|\frac{\partial^{j+k} u(x,t)}{\partial x^j \partial t^k}\right| \leq c_{jk}(1+x^2+t^2)^{\frac{\alpha_{jk}}{2}}, \quad (x,t) \in R^2{}_+, \quad k \geq 0, \quad j \geq 0, \tag{1.13}$$

where c_{jk} and α_{jk} are some constants and $\alpha_{jk} < \alpha$.

Let $N_\alpha(R^1)$ denote the class of infinitely differentiable in R^1 functions $f(x)$ satisfying the conditions

$$\left|\frac{\partial^k f(x)}{\partial x^k}\right| \leq c_k(1+x^2)^{\frac{\alpha_k}{2}}, \quad x \in R^1, \quad k \geq 0,$$

where c_k and α_k are some constants and $\alpha_k < \alpha$.

A solution of the problem (1.1), (1.9), (1.10) is searched in the class of functions $M_0(R^2_+)$. We impose the following conditions on $b_k(x)$ and $f_k(x)$:
1) there exist non-zero limits of functions $b_k(x)$ as $x \to +\infty$ and $x \to -\infty$ and these limits are equal (denote this common limit by $b_k(\infty)$).
2) $b_k(x) - b_k(\infty) \in N_0(R^1)$ and $f_j(x) \in N_0(R^1)$, $k = r, \dots, q-1$, $j = 0, \dots, q-1$.

3) The indices of $b_k(x)$, $k = r, \dots, q-1$ on the X–axis are equal to zero.

Let $b(x)$ be a complex-valued function on the X–axis and $b(x) \neq 0$, $x \in R^1$. The index of function $f(x)$ on the X–axis is equal to the increment of the argument of $f(x)$, for x from $-\infty$ to $+\infty$, divided by 2π.

The main result of this chapter is the following theorem.

Theorem 1.1 *The problem (1.1), (1.9), (1.10) is uniquely solvable in the class $M_0(R^2_+)$.*

In the above mentioned papers [1] – [10] the general boundary value problems for regular equations (1.1) were investigated provided that the coefficients of boundary conditions are constant, which was essentially used for investigation of these problems.

The problem (1.1), (1.9), (1.10) for the elliptic equation

$$\sum_{k=0}^{n} A_k \frac{\partial^n u}{\partial x^k \partial t^{n-k}} = 0$$

was investigated in [11], [12], in the case where the functions $b_k(x)$ are constant.

The essentially new point in our approach as compared with previous papers is as follows:

1) We construct the class $M_0(R^2_+)$ of functions to be found, where the considered problem is uniquely solvable.

2) The solutions of Cauchy problem for equation (1.1) with special boundary data are used for investigation of the problem (1.1), (1.9), (1.10).

In chapters 2 and 3 there are other more efficient methods of resolution of the problem (1.1), (1.9), (1.10) for the case, where equation (1.1) is a product of first-order differential operators.

1.2 Some Auxiliary Propositions

Let $\lambda_1(\sigma), \dots, \lambda_n(\sigma)$ be the roots of equation (1.2) and

$$Re\lambda_1(\sigma) \leq Re\lambda_2(\sigma) \leq \dots \leq Re\lambda_n(\sigma).$$

Each root is counted according to its multiplicity. We assume that conditions (1.7) and (1.8) are satisfied. According to them, we have

$$Re\lambda_j(\sigma) \leq 0, \quad j = 1, \dots, r, \quad \sigma > 0, \tag{1.14}$$

$$Re\lambda_j(\sigma) \leq 0, \quad j = 1, \dots, q, \quad \sigma < 0, \tag{1.15}$$

$$Re\lambda_j(\sigma) > 0, \quad j = r+1, \dots, n, \quad \sigma > 0, \tag{1.16}$$

$$Re\lambda_j(\sigma) > 0, \quad j = q+1, \dots, n, \quad \sigma < 0. \tag{1.17}$$

Let

$$(\lambda - \lambda_1(\sigma)) \cdots (\lambda - \lambda_r(\sigma)) \equiv \lambda^r + \alpha_1(\sigma)\lambda^{r-1} + \cdots + \alpha_r(\sigma),$$
$$(\lambda - \lambda_1(\sigma)) \cdots (\lambda - \lambda_q(\sigma)) \equiv \lambda^q + \beta_1(\sigma)\lambda^{q-1} + \cdots + \beta_q(\sigma).$$

Consider the ordinary differential equations

$$\frac{d^r y}{dt^r} + \alpha_1(\sigma)\frac{d^{r-1}y}{dt^{r-1}} + \cdots + \alpha_r(\sigma)y = 0, \quad \sigma > 0, \ t \geq 0, \tag{1.18}$$
$$\frac{d^q y}{dt^q} + \beta_1(\sigma)\frac{d^{q-1}y}{dt^{q-1}} + \cdots + \beta_q(\sigma)y = 0, \quad \sigma < 0, \ t \geq 0, \tag{1.19}$$

depending on the parameter σ. Let $y_{1k}(\sigma, t)$ $(k = 0, \ldots, q-1)$ and $y_{2l}(\sigma, t)$ $(l = 0, \ldots, r-1)$ be the solutions of equations (1.18) and (1.19) respectively, satisfying the Cauchy boundary conditions

$$\frac{d^j y_{1k}(\sigma, 0)}{dt^j} = \begin{cases} 1, & \text{for } j = k \\ 0, & \text{for } j \neq k, \ j = 0, \ldots, r-1, \end{cases} \tag{1.20}$$
$$\frac{d^j y_{2l}(\sigma, 0)}{dt^j} = \begin{cases} 1, & \text{for } j = l \\ 0, & \text{for } j \neq l, \ j = 0, \ldots, q-1. \end{cases} \tag{1.21}$$

Let the roots $\lambda_1(\sigma), \ldots, \lambda_r(\sigma)$ satisfy the inequalities (1.14) – (1.17).

Lemma 1.1 *The functions $\alpha_1(\sigma), \ldots, \alpha_r(\sigma)$ are infinitely differentiable at any point $\sigma \in R$, $\sigma > 0$ and satisfy the estimates*

$$| \alpha_m(\sigma) | \leq c_0, \quad m = 1, \ldots, r, \tag{1.22}$$
$$\left| \frac{d^k \alpha_m(\sigma)}{d\sigma^k} \right| \leq c_k \, | \sigma |^{\gamma - k} \, (1 + | \sigma |)^{l_k}, \tag{1.23}$$

$\sigma > 0$, $m = 1, \ldots, r$; $k = 1, 2, \ldots,$ where c_k, l_k and γ are some positive constants. A similar assertion is true for the functions $\beta_1(\sigma), \ldots, \beta_r(\sigma)$ for $\sigma < 0$.

Lemma 1.2 *The functions $y_{1k}(t, \sigma)$ $(y_{2l}(t, \sigma))$ are continuous for $t \geq 0$, $\sigma \geq 0$ $(t \geq 0, \ \sigma \leq 0)$ and are infinitely differentiable for $t \geq 0$, $\sigma > 0$ $(t \geq 0, \sigma < 0)$ and satisfy the estimates*

$$\left| \frac{\partial^{l+m} y_{1k}(t, \sigma)}{\partial t^l \partial \sigma^m} \right| \leq c_{lm} (1 + t)^{l_m} (1 + | \sigma |)^{\delta_{lm}} \, | \sigma |^{\gamma - m} \tag{1.24}$$

for $t > 0$, $\sigma > 0$, where $k = 0, \ldots, q-1$, $m, l = 0, 1, \ldots$; c_{lm}, l_m, δ_{lm} and γ are some positive constants (the functions $y_{2l}(t, \sigma)$ satisfy the same estimates for $t > 0$, $\sigma < 0$).

The proofs of Lemmas 1.1 and 1.2 can be found in [10], [13].

Let S denote the class of infinitely differentiable and rapidly decreasing functions in R^1, i.e. S is the set of functions $\varphi(x)$ satisfying

$$\lim_{|x|\to+\infty} x^m \varphi^{(k)}(x) = 0, \quad m, k = 0, 1, 2, \ldots. \tag{1.25}$$

For $\varphi(x) \in S$, we denote by $F(\varphi)$ and $F^{-1}(\varphi)$ the direct and inverse Fourier transforms of the function $\varphi(x)$, i.e.

$$F(\varphi) = \frac{1}{\sqrt{2\pi}} \int_{-\infty}^{+\infty} \varphi(\zeta) e^{i\zeta x} d\zeta, \quad F^{-1}(\varphi) = \frac{1}{\sqrt{2\pi}} \int_{-\infty}^{+\infty} \varphi(\zeta) e^{-i\zeta x} d\zeta. \tag{1.26}$$

These functions also belong to S.

Any function $f(x) \in N_0(R^1)$ generates a generalized function f by

$$(f, \varphi) = \int_{-\infty}^{+\infty} f(x)\varphi(x) dx, \quad \varphi(x) \in S. \tag{1.27}$$

In this case we will write $f \in N_0$.

The direct and inverse Fourier transforms ($F(f)$ and $F^{-1}(f)$) of a generalized function $f \in N_0$ are defined by

$$(F(f), \varphi) = (f, F(\varphi)), \quad (F^{-1}(f), \varphi) = (f, F^{-1}(\varphi)), \quad \varphi \in S.$$

If $\psi(x)$ is a given bounded function in R^1 then we define

$$(\psi f, \varphi) = \int_{-\infty}^{+\infty} f(x)\psi(x)\varphi(x) dx \equiv (f, \psi\varphi), \quad \varphi \in S.$$

Lemma 1.3 *Let a generalized function $f \in N_0$ and let ψ be an infinitely differentiable function satisfying the condition*

$$|\psi(x)| \le \frac{c}{1 + |x|^2}, \quad |\psi'(x)| \le \frac{c}{(1 + |x|^2)|x|^\gamma}, \quad x \in R^1, x \ne 0, \tag{1.28}$$

where c and γ $(0 \le \gamma < 1)$ are some positive constants.
Then the generalized function $g = F^{-1}(\psi(x)F(f))$ belongs to the class N_0 and the corresponding function $g(x)$ is defined by the formula

$$g(x) = \int_{-\infty}^{+\infty} f(\zeta)\omega(x - \zeta) d\zeta, \tag{1.29}$$

where

$$\omega(\sigma) = \frac{1}{2\pi}\int\limits_{-\infty}^{+\infty}\psi(\zeta)e^{-i\zeta\sigma}d\zeta. \tag{1.30}$$

Proof. It follows from (1.28) that

$$|\,\omega(\sigma)\,| \leq c(1+|\,\sigma\,|)^{-1}, \quad \sigma \in R^1, \tag{1.31}$$

where c is some constant.

It is easy to check that

$$\frac{1}{1+|\,(x-\zeta)\,|} \leq \frac{1+|\,\zeta\,|}{1+|\,x\,|}, \quad x \in R^1, \quad \zeta \in R^1. \tag{1.32}$$

Hence, from (1.31) we get

$$|\,\omega(x-\zeta)\,| \leq \frac{1+|\,\zeta\,|}{1+|\,x\,|}. \tag{1.33}$$

Let $g = F^{-1}(\psi(x)F(f))$, then

$$\begin{aligned}(g,\varphi) &= (F^{-1}(\psi(x)F(f)),\varphi) = (\psi(x)F(f),F^{-1}(\varphi))\\ &= (F(f),\psi(x)F^{-1}(\varphi)) = (f,F(\psi(x)F^{-1}(\varphi)))\\ &= \int\limits_{-\infty}^{+\infty} f(x)F(\psi(x)F^{-1}(\varphi))dx,\end{aligned} \tag{1.34}$$

where

$$F(\psi(x)F^{-1}(\varphi)) = \frac{1}{2\pi}\int\limits_{-\infty}^{+\infty}\int\limits_{-\infty}^{+\infty}\psi(\sigma)\varphi(\zeta)e^{-i\sigma(x-\zeta)}d\zeta d\sigma. \tag{1.35}$$

Since $\psi(x)$ satisfies the first inequality of (1.28), according to Fubini's theorem, one can change the order of integration in (1.35) to obtain

$$F(\psi(x)F^{-1}(\varphi)) = \int\limits_{-\infty}^{+\infty}\omega(\zeta-x)\varphi(\zeta)d\zeta. \tag{1.36}$$

From (1.34) and (1.36) we have

$$(g,\varphi) = \int\limits_{-\infty}^{+\infty}\int\limits_{-\infty}^{+\infty} f(x)\omega(\zeta-x)\varphi(\zeta)d\zeta dx. \tag{1.37}$$

Since $f(x) \in N_0(R^1)$ and $\omega(\zeta - x)$ satisfies the estimate (1.33), according to Fubini's theorem, one can change the order of integration in (1.37). Hence

$$(g, \varphi) = \int_{-\infty}^{+\infty}\int_{-\infty}^{+\infty} \varphi(\zeta) f(x)\omega(\zeta - x) dx d\zeta. \tag{1.38}$$

Using (1.29), the equality (1.38) may be rewritten as

$$(g, \varphi) = \int_{-\infty}^{+\infty} \varphi(\zeta) g(\zeta) d\zeta. \tag{1.39}$$

It is clear that $g(x)$ can be represented as (by change of variable $x - \eta = \zeta$):

$$g(x) = \int_{-\infty}^{+\infty} f(x - \zeta)\omega(\zeta) d\zeta. \tag{1.40}$$

Since $f \in N_0(R^1)$ and $\omega(\sigma)$ satisfies (1.31), we have $g(x) \in N_0(R^1)$.

Lemma 1.3 is proved.

Note 1.1. The Fourier transform of a function belonging to the class $N_0(R^1)$ in the ordinary sense may fail to exist, but it always exists as a generalized function. For this reason we use the generalized functions in our proofs.

Consider the Cauchy type integral

$$u(z) = \frac{1}{2\pi i}\int_{-\infty}^{+\infty} \frac{f(\zeta) d\zeta}{\zeta - z}, \quad z = x + iy, \quad x \in R^1, \quad y > 0, \tag{1.41}$$

where $f(x)$ is a given function belonging to the class $N_0(R^1)$.

Lemma 1.4 *The function $u(z)$ defined by (1.41) is analytic on $z = x + iy$ and belongs to the class $N_0(R^2_+)$.*

This lemma can be proved using the method, described in [14] (pp. 76–83). Denote by $u^+(x)$ the boundary value of a function $u(z)$ as $z \to x$, $Im z > 0$. By Sokhotzky-Plemelj formula (cf. [15], p. 66) we have

$$u^+(x) = \frac{f(x)}{2} + \frac{1}{2\pi i}\int_{-\infty}^{+\infty} \frac{f(\zeta) d\zeta}{\zeta - x}, \tag{1.42}$$

where the singular integral is a Cauchy principal value. Since $u(z) \in N_0(R^2_+)$, then $u^+(x) \in N_0(R^1)$.

Let u^+ be the generalized function corresponding to the function $u^+(x)$, i.e.

$$(u^+, \varphi) = \int_{-\infty}^{+\infty} u^+(x)\varphi(x)dx. \tag{1.43}$$

Let

$$\eta(x) = 0 \quad \text{for} \quad x > 0 \quad \text{and} \quad \eta(x) = 1 \quad \text{for} \quad x \leq 0. \tag{1.44}$$

Lemma 1.5 *Let $f(x) \in N_0(R^1)$ and let f be the corresponding generalized function, then*

$$\eta(x)F(f) \;=\; F(u^+). \tag{1.45}$$

Proof. Let $y > 0$ be a fixed number, and let g_0 be the generalized function

$$g_0 \;=\; F^{-1}(\eta(x)e^{yx}F(f)). \tag{1.46}$$

From Lemma 1.3 it follows that the generalized function $g_0 \in N_0$ and its corresponding function is defined by (cf. (1.41))

$$(g_0, \varphi) \;=\; \int_{-\infty}^{+\infty} u(x+iy)\varphi(x)dx, \quad \varphi \in S. \tag{1.47}$$

Hence

$$(F(g_0), \varphi) \equiv (g_0, F(\varphi)) \;=\; \int_{-\infty}^{+\infty} u(x+iy)F(\varphi)dx, \quad \varphi \in S, \tag{1.48}$$

where $F(\varphi)$ is the Fourier transform of a function $\varphi(x)$ (cf. (1.26)). From (1.46) we have

$$F(g_0) \;=\; \eta(x)e^{yx}F(f). \tag{1.49}$$

Substituting $F(g_0)$ from (1.49) into (1.48) we get

$$(\eta(x)e^{yx}F(f), \varphi) \;=\; \int_{-\infty}^{+\infty} u(x+iy)F(\varphi)dx, \quad \varphi \in S. \tag{1.50}$$

By the definition of Fourier transform of a generalized function F, we have

$$\begin{aligned}(\eta(x)e^{yx}F(f), \varphi) &= (F(f), \eta(x)e^{yx}\varphi) \;=\; (f, F(\eta(x)e^{yx}\varphi(x))) \\ &= \int_{-\infty}^{+\infty} f(x)F(\eta(x)e^{yx}\varphi(x))dx, \end{aligned} \tag{1.51}$$

where

$$F(\eta(x)e^{yx}\varphi(x)) = \frac{1}{\sqrt{2\pi}} \int_{-\infty}^{+0} e^{y\zeta}\varphi(\zeta)e^{i\zeta x}d\zeta. \tag{1.52}$$

Integration by parts the right-hand side of (1.51) gives

$$\left| \int_{-\infty}^{0} e^{y\zeta}\varphi(\zeta)e^{i\zeta x}d\zeta \right| \leq \frac{c}{1+ \mid x \mid}, \tag{1.53}$$

where c is a constant, independent of x and y.

Passing to the limit in the equality (1.51) as $y \to +0$, we get

$$\lim_{y\to+0}(\eta(x)e^{yx}F(f),\varphi) = \frac{1}{\sqrt{2\pi}} \int_{-\infty}^{+\infty}\int_{-\infty}^{0} f(x)\varphi(\zeta)e^{i\zeta x}d\zeta dx. \tag{1.54}$$

Next, passing to the limit on both sides of (1.50) as $y \to +0$ and taking into account the equality (1.54) we have

$$\frac{1}{\sqrt{2\pi}} \int_{-\infty}^{+\infty}\int_{-\infty}^{0} f(x)\varphi(\zeta)e^{i\zeta x}d\zeta dx = \int_{-\infty}^{+\infty} u^+(x)F(\varphi)dx. \tag{1.55}$$

Observe that (1.55) is an abridged version of the equality (1.45).

Lemma 1.5 is proved.

Lemma 1.6 *Let $f(x) \in N_0(R^1)$ and let f be the corresponding generalized function. Then $f(x)$ is the boundary value of an analytic with respect to $z = x + iy$ function $f(z)$ from the class $N_0(R^2_+)$ if and only if*

$$(1 - \eta(x))F(f) = 0. \tag{1.56}$$

Proof. Let $f(z)$ be analytic in the half-plane $Imz > 0$ and $f(z) \in N_0(R^2_+)$. Then, by Cauchy formula

$$f(z) = u(z), \quad f(x) = u^+(x), \tag{1.57}$$

where $u(z)$ and $u^+(x)$ are defined by (1.41) and (1.42).

Let u^+ be the generalized function corresponding to $u^+(x)$. From the second equality of (1.57) we have $u^+ = f$. This and (1.45) imply the equality (1.56).

Now let the equality (1.56) be satisfied. Then from (1.45) and (1.56) we have $F(f) = F(u^+)$. So $f(x) = u^+(x)$. Hence f is the boundary value of the analytic function $u(z)$ defined by formula (1.41). Since $f \in N_0(R^1)$, then by Lemma 1.3 we obtain $u(z) \in N_0(R^2_+)$.

Lemma 1.6 is proved.

In a similar way we get

Lemma 1.7 *A function $f(x)$ is the boundary value of an analytic in the half-plane $Imz < 0$ function $f(z) \in N_0(R^2_-)$ if and only if $f(x) \in N_0(R^1)$ and $\eta(x)F(f) = 0$.*

Let $\varphi(x) \in S$ and let Ω_0 be the set of points $x \in R^1$ such that $\varphi(x) \neq 0$. The closure Ω_0 is called the support of function φ and is denoted by $supp\varphi$.

Let Ω be an open subset of R^1 and let f and g be the generalized functions from class N_0. We say that these generalized functions are equal in Ω, if

$$f = g, \quad x \in \Omega, \quad \text{i.e.} \quad (f, \varphi) = (g, \varphi) \quad \forall \varphi \in S, \quad \text{supp}\varphi \subset \Omega.$$

Lemma 1.8 *If $f(x) \in N_0(R^1)$, $g(x) \in N_0(R^1)$ and*

$$F(f) = F(g) \quad for \quad x \in R^1 \setminus 0, \tag{1.58}$$

then $f(x) = g(x)$.

Proof. From (1.58) it follows that the generalized function $F(f - g)$ is concentrated at the point $x = 0$. So (see [3]), we have

$$F(f-g) = \sum_{k=0}^{m} c_k \delta^{(k)}, \tag{1.59}$$

where $\delta^{(k)}$ is the k-th derivative of Dirac's delta function δ $((\delta, \varphi) = \varphi(0))$ and c_k are some constants. It follows from (1.59) that $f(x) - g(x)$ is a polynomial on x. Because $f(x) - g(x) \in N_0(R^1)$, then $f(x) - g(x) \equiv 0$.

Lemma 1.8 is proved.

1.3 Cauchy Problem for Equation (1.1) in the Class of Analytic Functions

The results of this section are essentially used for investigation of the problem (1.1), (1.9), (1.10) and for the proof of Theorem 1.1.

Let $u(x,t) \in M_0(R^2_+)$. For every fixed $t \geq 0$ to the function $u(x,t)$ there corresponds a generalized function $u(t)$ in the following way

$$(u(t), \varphi) = \int_{-\infty}^{+\infty} u(x,t)\varphi(x)dx, \quad \varphi \in S. \tag{1.60}$$

The generalized function $u(t)$ depends on t as a parameter. Denote by $F_x(u(t))$ the Fourier transform of the generalized function $u(t)$ at a fixed value of t, i.e.

$$(F_x(u(t)), \varphi) = (u(t), F(\varphi)) = \int_{-\infty}^{+\infty} u(x,t)F(\varphi)dx, \quad \varphi \in S, \tag{1.61}$$

where $F(\varphi)$ is defined by (1.26).

For $u(x,t) \in M_0(R^2_+)$ we denote by $u(t)$ and $F_x(u(t))$ the generalized functions, which depend on t as a parameter and defined on S by (1.60) and (1.61), respectively.

Definition 1.4. We define $M_0^+(R^2_+)$ as the class of functions $u(x,t) \in M_0(R^2_+)$, which for any fixed $t \geq 0$ satisfy the equality

$$(1-\eta(x))F_x(u(t)) = 0, \tag{1.62}$$

where $\eta(x)$ is defined by (1.44).

Definition 1.5. In a similar way, we define $M_0^-(R^2_+)$ as the class of functions $u(x,t) \in M_0(R^2_+)$, which for any fixed $t \geq 0$ satisfies the equality

$$\eta(x)F_x(u(t)) = 0. \tag{1.63}$$

If $u(x,t) \in M_0(R^2_+)$ and the condition (1.62) is fulfilled, then according to Lemma 1.5, the function $u(x,t)$ for any fixed $t \geq 0$ is the boundary value of some analytic on $z = x+iy$ function $u(z,t)$ in $Imz > 0$. Similarly, the condition (1.63) means that for any fixed $t \geq 0$ the function $u(x,t)$ is the boundary value of some analytic on $z = x+iy$ function $u(z,t)$ in $Imz < 0$.

Now we consider equation (1.1) with boundary conditions

$$\frac{\partial^k u(x,0)}{\partial t^k} = g_k(x), \quad x \in R^1, \quad k = 0,1,\ldots,q-1, \tag{1.64}$$

where $g_k(x) \in N_0(R^1)$ and

$$(1-\eta(x))F(g_k) = 0, \quad k = 0,1,\ldots,q-1. \tag{1.65}$$

Theorem 1.2 *The problem (1.1), (1.64) is uniquely solvable in the class $M_0^+(R^2_+)$.*

Proof. We pass to the Fourier transforms in equation (1.1) and in the boundary conditions (1.64) with respect to variable x to obtain

$$\frac{\partial^n v(t)}{\partial t^n} + \sum_{k=0}^{n-1} P_k(x)\frac{\partial^k v(t)}{\partial t^k} = 0, \quad t > 0, \tag{1.66}$$

$$\frac{\partial^k v(0)}{\partial t^k} = F(g_k), \quad k = 0,1,\ldots,q-1, \tag{1.67}$$

where

$$v(t) = F_x(u(t)). \tag{1.68}$$

Here

$$\left(\frac{d^k v(t)}{dt^k}, \varphi\right) \equiv \frac{d^k}{dt^k}(v(t), \varphi(x)), \quad \varphi \in S,$$
$$\left(P_k(x)\frac{d^k v(t)}{dt^k}, \varphi(x)\right) \equiv \left(\frac{d^k v(t)}{dt^k}, P_k(x)\varphi(x)\right), \quad \varphi(x) \in S.$$

Since $u(x,t) \in M_0^+(R_+^2)$, then by the definition of the class $M_0^+(R_+^2)$, we have

$$(1-\eta(x))v(t) = 0, \quad t \geq 0. \tag{1.69}$$

Let

$$(\lambda - \lambda_1(\sigma))\cdots(\lambda - \lambda_q(\sigma)) = \lambda^q + \beta_1(\sigma)\lambda^{q-1} + \cdots + \beta_q(\sigma), \tag{1.70}$$

and

$$(\lambda - \lambda_{q+1}(\sigma))\cdots(\lambda - \lambda_n(\sigma)) = \lambda^{n-q} + \gamma_1(\sigma)\lambda^{n-q-1} + \cdots + \gamma_{n-q}(\sigma). \tag{1.71}$$

By Lemma 1.1 the functions $\beta_1(\sigma), \ldots, \beta_q(\sigma)$ and $\gamma_1(\sigma), \ldots, \gamma_{n-q}(\sigma)$ are infinitely differentiable in $\sigma \in (-\infty, 0)$ and satisfy the same estimates as $\alpha_m(\sigma)$ in Lemma 1.1 (cf. (1.22) and (1.23)).

Denote by L_1 and L_2 the differential operators

$$L_1 v \equiv \frac{d^{n-q}v}{dt^{n-q}} + \gamma_1(x)\frac{d^{n-q-1}v}{dt^{n-q-1}} + \cdots + \gamma_{n-q}(x)v,$$
$$L_2 v \equiv \frac{d^q v}{dt^q} + \beta_1(x)\frac{d^{q-1}v}{dt^{q-1}} + \cdots + \beta_q(x)v.$$

Equation (1.66) in the interval $(-\infty, 0)$ may be rewritten as

$$L_1 L_2 v = 0, \quad t > 0, \quad x \in (-\infty, 0). \tag{1.72}$$

First we prove the existence of a solution for the problem (1.66), (1.67) satisfying the condition (1.69).

To this end let us consider the equation

$$L_2 v = 0, \quad t > 0, \quad x \in (-\infty, 0) \tag{1.73}$$

with the boundary conditions (1.67).

Let $y_{2,0}(\sigma,t), \ldots, y_{2,q-1}(\sigma,t)$ be the fundamental system of solutions of (1.19) for $\sigma \leq 0$ in the class of ordinary functions satisfying (1.21) and

$$y_{2j}(\sigma,t) = 0 \quad \textit{for} \quad \sigma > 0, \quad t \geq 0, \quad j = 0, \ldots, q-1. \tag{1.74}$$

Consider the following generalized function $v(t)$:

$$v(t) = \sum_{j=0}^{q-1} y_{2j}(x,t)F(g_k). \tag{1.75}$$

Since $y_{2j}(\sigma, t)$ $(j = 0, \ldots, m-1)$ is the fundamental system of solutions of (1.19), for $t > 0$ and $x \in (-\infty, 0)$, $v(t)$ satisfies (1.73). Hence for $t > 0, x \in (-\infty, 0)$, $v(t)$ also satisfies equation (1.72). It follows from (1.74) that

$$(1 - \eta(x))v(t) = 0 \quad t \geq 0. \tag{1.76}$$

Hence for $x \in (-\infty, 0) \cup (0, \infty)$ we have

$$\frac{d^n v(t)}{dt^n} + \sum_{k=0}^{n-1} P_k(x) \frac{d^k v(t)}{dt^k} = 0, \quad t > 0. \tag{1.77}$$

Using (1.21) and (1.65) in a similar way for $x \in (-\infty, 0) \cup (0, \infty)$ we obtain

$$\frac{d^k v(0)}{dt^k} = F(g_k), \quad k = 0, \ldots, q-1, \quad t > 0. \tag{1.78}$$

Let $u(x,t)$ be the inverse Fourier transform of a generalized function $v(t)$ with respect to variable t. We prove that $u(x,t) \in M_0(R^2_+)$ and it is a solution of the Cauchy problem (1.1), (1.64) in the class $M_0(R^2_+)$.

It is known that

$$F(g^{(k)}(x)) = (-ix)^k F(g(x)).$$

Denote

$$\mu_k(x) = (I - \frac{\partial}{\partial x})^m g_k(x), \tag{1.79}$$

where I is the identity operator and m is a sufficiently large positive integer. It is clear that

$$F(\mu_k) = (1 + ix)^m F(g_k). \tag{1.80}$$

Since $g_k \in N_0(R^1)$, then $\mu_k \in N_0(R)$. Substituting $F(g_k)$ from (1.80) into (1.75) we get

$$v(t) = \sum_{j=0}^{q-1} (1 + ix)^{-m} y_{kj}(x,t) F(\mu_k). \tag{1.81}$$

From (1.81) we obtain

$$u(x,t) = F_x{}^{-1}(v(t)) = \sum_{j=0}^{q-1} F_x{}^{-1}((1 + ix)^{-m} y_{kj}(x,t)F(\mu_k)). \tag{1.82}$$

By Lemmas 1.2 and 1.3 the function $u(x,t)$ belongs to $M_0(R^2_+)$ for large values of m and is defined by the formula

$$u(x,t) = \sum_{j=0}^{q-1} \int_{-\infty}^{+\infty} \mu_k(x-\zeta)\omega_k(\zeta,t)d\zeta, \qquad (1.83)$$

where

$$\omega_k(\sigma,t) = \frac{1}{2\pi} \int_{-\infty}^{+\infty} (1+i\zeta)^{-m} y_k(\zeta,t)e^{-i\zeta\sigma}d\zeta. \qquad (1.84)$$

Since $u(x,t) \in M_0(R^2_+)$, $v(t) = F_x(u(x,t))$, by (1.76) the function $u(x,t)$ belongs to $M_0^+(R^2_+)$ (cf. Definition 1.4). Denote by $w(x,t)$ the left-hand side of (1.1), where $u(x,t)$ is defined by (1.83). The equality (1.77) takes the form

$$F_x(w(x,t)) = 0, \quad t > 0, \quad x \in (-\infty,0) \cup (0,\infty). \qquad (1.85)$$

So the generalized function $F_x(w(x,t))$ is concentrated at the point $x = 0$. Hence

$$w(x,t) = \sum_{k=0}^{l} C_k(t)x^k, \qquad (1.86)$$

where l is some positive integer. Since $w(x,t) \in M_0(R^2_+)$, then $c_k(t) \equiv 0$ and $w(x,t) \equiv 0$. Thus $u(x,t)$ satisfies equation (1.1). Further, $u(x,0) \in N_0(R^1)$, hence from (1.78) and Lemma 1.8 it follows that $u(x,t)$ satisfies the Cauchy conditions (1.64).

Thus, the existence of a solution of the problem (1.1), (1.64) is proved.

Let $u(x,t)$ be any solution of the problem (1.1), (1.64) at $g_k(x) = 0$, $k = 0,1,\dots,q-1$. Then $v(t) = F_x(u(x,t))$ satisfies equation (1.72). Denote $L_2 v = \omega(t)$. Then equation (1.72) takes the form

$$L_1\omega(t) = 0, \quad t > 0, \quad x \in (-\infty,0). \qquad (1.87)$$

Since the roots $\lambda_{q+1}(\sigma),\dots,\lambda_n(\sigma)$ of the characteristic equation (1.87) satisfy the inequality (1.17), for $x \in (-\infty,0)$ a solution of equation (1.87) is equal to zero (cf. [3]), i.e.

$$L_2 v = 0, t > 0, x \in (-\infty,0). \qquad (1.88)$$

Hence $v(t)$ satisfies equation (1.88) and the homogeneous boundary conditions (1.67) ($F(g_k) \equiv 0$, $k = 0,\dots,q-1$). From uniqueness of a solution of the problem (1.88), (1.67) it follows that (cf. [3])

$$v(t) = 0, \quad t > 0, \quad x \in (-\infty,0). \qquad (1.89)$$

From (1.69) and (1.89) we get

$$v(t) = 0, \quad t > 0, \quad x \in (-\infty, 0) \cup (0, \infty).$$

Hence by Lemma 1.8 we have $v(t) \equiv 0$ and $u(x,t) \equiv 0$.

Theorem 1.2 is proved.

Consider now equation (1.1) with the boundary conditions

$$\frac{\partial^k u(x,0)}{\partial t^k} = g_k(x), \quad x \in R^1, \quad k = 0, \ldots, r-1, \tag{1.90}$$

where $g_k(x) \in N_0(R^1)$ and

$$\eta(x) F(g_k) = 0, \quad k = 0, \ldots, r-1.$$

Theorem 1.3 *The problem (1.1), (1.90) is uniquely solvable in the class* $M_0^-(R_+^2)$.

The proof is similar to that of Theorem 1.2.

1.4 Dirichlet Type Problem for Equation (1.1)

Let the roots $\lambda_1(\sigma), \ldots, \lambda_n(\sigma)$ of the characteristic equation (1.1) satisfy the conditions (1.14) – (1.17). To prove Theorem 1.1 we need the following

Lemma 1.9 *Let $u(x,t)$ be a solution of equation (1.1) belonging to* $M_0(R_+^2)$, *then it admits the representation*

$$u(x,t) = \omega^+(x,t) - \omega^-(x,t), \tag{1.91}$$

where $\omega^+(x,t)$ and $\omega^-(x,t)$ are the solutions of equation (1.1) satisfying $\omega^+(x,t) \in M_0^+(R_+^2)$ *and* $\omega^-(x,t) \in M_0^-(R_+^2)$.

Proof. Let us consider the function

$$\omega(x+iy,t) = \frac{1}{2\pi i} \int_{-\infty}^{+\infty} \frac{u(\zeta,t)d\zeta}{\zeta - x - iy}, \quad y \neq 0, \tag{1.92}$$

where $u(x,t)$ is a solution of (1.1) belonging to $M_0(R_+^2)$.

By Sokhotzky-Plemelj formula (see [15]) we have

$$\omega^+(x,t) \equiv \lim_{y \to +0} \omega(x+iy,t) = \frac{1}{2}u(x,t) + \frac{1}{2\pi i} \int_{-\infty}^{+\infty} \frac{u(\zeta,t)d\zeta}{\zeta - x}, \tag{1.93}$$

$$\omega^-(x,t) \equiv \lim_{y\to -0} \omega(x+it,t) = -\frac{1}{2}u(x,t) + \frac{1}{2\pi i}\int_{-\infty}^{+\infty} \frac{u(\zeta,t)d\zeta}{\zeta - x}. \qquad (1.94)$$

The singular integrals in (1.93) and (1.94) are Cauchy principal value.

From (1.93) and (1.94) we have

$$u(x,t) = \omega^+(x,t) - \omega^-(x,t). \qquad (1.95)$$

Since $u(x,t) \in M_0(R^2_+)$, from Lemmas 1.4, 1.6 and 1.7 it follows that

$$\omega^+(x,t) \in M_0^+(R^2_+) \quad \text{and} \quad \omega^-(x,t) \in M_0^-(R^2_-).$$

From (1.92) we get

$$\frac{\partial^{k+j}\omega(x+iy,t)}{\partial x^k \partial t^j} = \frac{1}{2\pi i}\int_{-\infty}^{+\infty} \frac{\partial^{k+j}u(x,\zeta)}{\partial t^k \partial \zeta^j}\frac{d\zeta}{\zeta - x - iy}, \quad y \neq 0. \qquad (1.96)$$

From (1.96) it follows that $\omega(x+iy,t)$ also satisfies equation (1.1), i.e.

$$\frac{\partial^n \omega(x+iy,t)}{\partial t^n} + \sum_{k=0}^{n-1} \frac{\partial^k}{\partial t^k}(P_k(i\frac{\partial}{\partial x})\omega(x+iy,t)) = 0, \qquad (1.97)$$

where $t > 0$, $x \in R^1$, and $y \neq 0$.

Passing to the limit in (1.97) as $y \to +0$ and $y \to -0$ we conclude that $\omega^+(x,t)$ and $\omega^-(x,t)$ are solutions of equation (1.1).

Lemma 1.9 is proved.

Proof of Theorem 1.1. By Lemma 1.9 a solution $u(x,t)$ of equation (1.1) belonging to the class $M_0(R^2_+)$ can be represented as in (1.91). Substituting $u(x,t)$ from (1.93) into (1.9) and (1.10), for $x \in R^1$ we get

$$\frac{\partial^k \omega^+(x,0)}{\partial t^k} - \frac{\partial^k \omega^-(x,0)}{\partial t^k} = f_k(x), \qquad k = 0,\dots,r-1, \qquad (1.98)$$

$$Re\left[b_k(x)(\frac{\partial^k \omega^+(x,0)}{\partial t^k} - \frac{\partial^k \omega^-(x,0)}{\partial t^k})\right] = f_k(x), \; k = r,\dots,q-1. \quad (1.99)$$

Since $\omega^+(x,t) \in M_0^+(R^2_+)$ and $\omega^-(x,t) \in M_0^-(R^2_-)$, by the definition of these spaces we have

$$\frac{\partial^k \omega^+(x,0)}{\partial t^k} \equiv \varphi_k(x), \quad k = 0,\dots,r-1, \qquad (1.100)$$

$$\frac{\partial^j \omega^-(x,0)}{\partial t^j} \equiv \psi_j(x), \quad j = 0,\dots,q-1, \qquad (1.101)$$

where $z = x + iy$, $\varphi_k(z) \in N_0(R^2_+)$, $\psi_j(z) \in N_0(R^2_-)$ and $\varphi_k(z)$ and $\psi_j(z)$ are analytic functions in R^2_+ and R^2_-, respectively. Here R^2_+ and R^2_- denote the half-planes $Imz > 0$ and $Imz < 0$, respectively.

Substituting the values of

$$\frac{\partial^k \omega^+(x,0)}{\partial t^k} \quad \text{and} \quad \frac{\partial^j \omega^-(x,0)}{\partial t^j}$$

from (1.100) and (1.101) into the condition (1.98), we get

$$\varphi_k(x) - \psi_k(x) = f_k(x), \quad x \in R^1, \quad k = 0, \dots, r-1, \tag{1.102}$$
$$Re[b_k(x)\psi_k(x)] = g_k(x), \quad x \in R^1, \quad k = r, \dots, q-1, \tag{1.103}$$

where

$$g_k(x) = -f_k(x) + Re\left[b_k(x)\frac{\partial^k \omega^+(x,0)}{\partial t^k}\right]. \tag{1.104}$$

So we get the conjugate problem (1.102) to determine the analytic functions $\varphi_k(z)$ and $\psi_k(z)$ $(k = 0, \dots, r-1)$, which is uniquely solvable and the solution is defined by (cf. [15])

$$\varphi_k(z) = \frac{1}{2\pi i}\int_{-\infty}^{+\infty} \frac{f_k(t)dt}{t-z}, \quad Imz > 0, \quad k = 0, \dots, r-1, \tag{1.105}$$

$$\psi_k(z) = \frac{1}{2\pi i}\int_{-\infty}^{+\infty} \frac{f_k(t)dt}{t-z}, \quad Imz < 0, \quad k = 0, \dots, r-1. \tag{1.106}$$

Since $\omega^+(x,t)$ and $\omega^-(x,t)$ are the solutions of (1.1), we have

$$\frac{\partial^n \omega^+(x,t)}{\partial t^n} + \sum_{k=0}^{n-1} \frac{\partial^k}{\partial t^k} P(i\frac{\partial}{\partial x})\omega^+(x,t) = 0, \quad t > 0, \ x \in R^1, \tag{1.107}$$

$$\frac{\partial^n \omega^-(x,t)}{\partial t^n} + \sum_{k=0}^{n-1} \frac{\partial^k}{\partial t^k} P(i\frac{\partial}{\partial x})\omega^-(x,t) = 0, \quad t > 0, \ x \in R^1. \tag{1.108}$$

So we get the Cauchy problem (1.107), (1.100) to determine the function $\omega^+(x,t)$ with $\varphi_k(z)$ defined by (1.105).

By Theorem 1.2 these problems are uniquely solvable in the class $M_0^+(R^2_+)$. Hence, the right-hand sides of the conditions (1.103) are well defined. Since the index of a function $b_k(x)$ on X–axis is equal to zero and $g_k(x) \in N_0(R^1)$, the analytic function $\psi_k(z)$ $(k = r, \dots, q-1)$ is uniquely defined by the condition (1.103) (cf. section 2.4).

So the right-hand side of (1.101) can be assumed to be well defined. Hence we get the Cauchy problem (1.108), (1.101) to determine the function $\omega^-(x,t)$, which is uniquely solvable. Substituting $\omega^+(x,t)$ and $\omega^-(x,t)$ into (1.91) we get a unique solution of the problem (1.1), (1.9), (1.10).

Theorem 1.1 is proved.

Concluding this section we consider more general boundary value problem for equation (1.1).

For the boundary conditions of $x \in R^1$ we take

$$\frac{\partial^k u(x,0)}{\partial t^k} = f_k(x), \qquad k = 0,\dots,r-1, \tag{1.109}$$

$$\sum_{l+j\le k} Re\left[b_{klj}(x)\frac{\partial^{l+j}u(x,0)}{\partial x^l \partial t^j}\right] = f_k(x), \quad k = r,\dots,q-1, \tag{1.110}$$

where $b_{klj}(x)$ and $f_k(x)$ are given functions in R^1 and

$$b_{k0k}(x) \neq 0, \quad x \in R^1, \quad k = r,\dots,q-1. \tag{1.111}$$

Without loss of generality we assume that $\mid b_{k0k}(x) \mid = 1$. The solutions of the problem (1.1), (1.109), (1.110) are searched in the class $M_0(R^2_+)$. The functions $b_{klj}(x)$ and $f_k(x)$ satisfy the following conditions:

1) There exist non-zero limits of functions $b_{k0k}(x)$ as $x \to +\infty$ and $x \to -\infty$ and they are equal (this limit will be denoted by $b_{k0k}(\infty)$);

2) For $l+j \le k,\ j \neq k,\ b_{klj}(x)$ and its derivatives are bounded and

$$b_{k0k}(x) - b_{k0k}(\infty) \in N_0(R^1), \quad k = r,\dots,q-1;$$
$$f_k(x) \in N_0(R^1), \quad k = 1,\dots,q-1;$$

3) For $k = r,\dots,q-1$ the indices of $b_{k0k}(x)$ on X-axis are equal to zero.

Theorem 1.4 *The problem (1.1), (1.109), (1.110) is uniquely solvable in the class $M_0(R^2_+)$.*

Proof. The proof of this theorem differs in minor details from that of Theorem 1.1. So we sketch the essential points and omit the details.

By Lemma 1.9 a solution $u(x,t)$ of equation (1.1) belonging to the class $M_0(R^2_+)$ can be represented as in (1.91). Since $\omega^+(x,t) \in M_0^+(R^2_+)$ and $\omega^-(x,t) \in M_0^-(R^2_-)$ one has the representations (1.100), (1.101).

Substituting $u(x,t)$ from (1.91) into (1.109) and using (1.100) and (1.101) we get the boundary condition (1.102) to determine the analytic functions $\varphi_k(x)$ and $\psi_k(z)$ $\quad (k = 0,\dots,r-1)$.

A solution of the conjugate problem (1.102) is defined by (1.105), (1.106). Hence, resolving the problem (1.107), (1.100) we define the function $\omega^+(x,t)$.

Substituting $u(x,t)$ from (1.91) into (1.110) and using (1.101), we get

$$\sum_{j=r}^{k}\sum_{l=0}^{k-j}[b_{klj}(x)\psi_j^{(l)}(x)] = g_k(x), \quad x \in R^1, \quad k = r, \dots, q-1, \tag{1.112}$$

where

$$\begin{aligned} g_k(x) &= \sum_{l+j\le k} Re\left[b_{klj}(x)\frac{\partial^{l+j}\omega^+(x,0)}{\partial x^l \partial t^j}\right] \\ &- \sum_{j=0}^{r-1}\sum_{l=0}^{k-j} b_{klj}(x)\psi_j^{(l)}(x) - f(x), \quad k = r, \dots, q-1. \end{aligned} \tag{1.113}$$

Since $\omega^+(x,t)$ and $\psi_j(x)$ $(j = 0, \dots, r-1)$ are well defined and $\omega^+(x,t) \in M_0^+(R_+^2)$, $\psi_j(x) \in N_0(R^1)$, then $g_k(x)$ $(k = r, \dots, q-1)$ are well defined functions from the class $N_0(R^1)$.

The boundary condition (1.112) for $k = r$ takes the form

$$Re[b_{r0r}(x)\psi_r(x)] = g_r(x), \quad x \in R^1. \tag{1.114}$$

Resolving the conjugate problem (1.114) we find an analytic in $Imz < 0$ function $\psi_r(z)$.

Substituting $\psi_r(z)$ into the boundary condition (1.112) at $k = r+1$, we get a similar problem for $\psi_{r+1}(z)$. Proceeding in the same way we define all the analytic in the half-plane $Imz < 0$ functions $\psi_r(z), \dots, \psi_{q-1}(z)$.

Recall that the functions $\psi_1(z), \dots, \psi_{r-1}(z)$ are defined by (1.106). So we get the Cauchy problem (1.108), (1.101) to determine $\omega^-(x,t)$, which is uniquely solvable (see Theorem 1.3).

Substituting $\omega^+(x,t)$ and $\omega^-(x,t)$ into (1.91) we find a unique solution of the problem (1.1), (1.109), (1.110).

Theorem 1.4 is proved.

Now let (1.7) and (1.8) be fulfilled everywhere, except a finite number of points:

$$\rho(\sigma) = r \quad \text{for} \quad \sigma > 0, \quad \sigma \ne \sigma_1, \dots, \sigma_m, \tag{1.115}$$

$$\rho(\sigma) = q \quad \text{for} \quad \sigma < 0, \quad \sigma \ne \sigma_{m+1}, \dots, \sigma_l, \tag{1.116}$$

where $\sigma_1, \dots, \sigma_m$ and $\sigma_{m+1}, \dots, \sigma_l$ are some points on real axis.

The results of this chapter remain valid if the conditions (1.7), (1.8) are replaced by the conditions (1.115), (1.116).

Assume that for a fixed point $x_0 \in R^1$ the conditions

$$\rho(\sigma) = r \quad \text{for} \quad \sigma > x_0, \quad \sigma \ne \sigma_1, \dots, \sigma_m, \tag{1.117}$$

$$\rho(\sigma) = q \quad \text{for} \quad \sigma < x_0, \quad \sigma \ne \sigma_{m+1}, \dots, \sigma_l, \tag{1.118}$$

are satisfied, where r and q $(r \leq q)$ are non-negative integers.

In this case the boundary conditions (1.10) have been replaced by the conditions

$$Re\left[b_k(x)\frac{\partial^k u(x,0)}{\partial t^k}e^{ix_0x}\right] = f_k(x), \quad k = r,\dots,q-1, \tag{1.119}$$

where $b_k(x)$ and $f_k(x)$ are similar to these in (1.10).

Theorem 1.5 *If the conditions (1.117), (1.118) are satisfied, then the problem (1.1), (1.9), (1.119) is uniquely solvable in the class* $M_0(R^2_+)$.

Proof. Setting

$$u(x,t) = v(x,t)e^{-ix_0x}, \tag{1.120}$$

we reduce the proof of Theorem 1.5 to that of Theorem 1.1.

1.5 Examples

To illustrate the results obtained in this chapter we consider two examples of non-regular differential equations.

Example 1.1. Consider the following n-th order homogeneous equation

$$\sum_{k=0}^{n} A_k \frac{\partial^n u}{\partial x^k \partial t^{n-k}} = 0, \quad t > 0, \quad x \in R^1, \tag{1.121}$$

where $A_0, A_1, \dots, A_n$ are given complex constants and $A_0 \neq 0$.

The characteristic equation (1.2) corresponding to (1.121) has the form

$$\sum_{k=0}^{n} A_k(-i\sigma)^k \lambda^{n-k} = 0, \quad \sigma \in R^1. \tag{1.122}$$

Setting in (1.122)

$$\lambda = -i\sigma\mu \tag{1.123}$$

we get

$$A_0\mu^n + A_1\mu^{n-1} + \dots + A_n = 0. \tag{1.124}$$

Equations (1.122) and (1.124) are considered with respect to complex variables λ and μ, respectively. In (1.122) the variable σ is a parameter.

Let r_1, r_2 and r_3 $(r_1 + r_2 + r_3 = n)$ be the numbers of roots of (1.124) with $Im\mu = 0$, $Im\mu > 0$ and $Im\mu < 0$, respectively. Let $\rho(\sigma)$ be the number of roots of the characteristic equation (1.122) with $Re\lambda \leq 0$ and $\sigma \in R^1$.

From the relation (1.123) it follows that

$$\rho(\sigma) = \begin{cases} r_1 + r_3, & \text{for } \sigma > 0, \\ r_1 + r_2, & \text{for } \sigma < 0, \\ n, & \text{for } \sigma = 0. \end{cases}$$

Let $r = \min(r_1 + r_2, r_1 + r_3)$ and $q = \max(r_1 + r_2, r_1 + r_3)$.

The boundary conditions (1.9) and (1.10) for (1.121) take the form

$$\frac{\partial^k u(x,0)}{\partial t^k} = f_k(x), \quad x \in R^1, \quad k = 0, \dots, r-1, \tag{1.125}$$

$$Re\left[b_k(x)\frac{\partial^k u(x,0)}{\partial t^k}\right] = f_k(x), \quad x \in R^1, \quad k = r, \dots, q-1, \tag{1.126}$$

where $f_k(x)$ and $b_k(x)$ are the same functions as in (1.9) and (1.10).

By Theorem 1.1 it follows that the problem (1.121), (1.125), (1.126) in the class $M_0(R^2_+)$ is uniquely solvable.

Example 1.2. Let L_1 and L_2 be differential operators defined by

$$L_1 u \equiv \frac{\partial u}{\partial x} + i\frac{\partial u}{\partial t} + au,$$

$$L_2 u \equiv \frac{\partial^2 u}{\partial x^2} + \frac{\partial^2 u}{\partial t^2} + b\frac{\partial u}{\partial x} - c^2 u,$$

where a, b and c are arbitrary real constants.

Consider the equation

$$L_1 L_2 u = 0, \quad t > 0, \quad x \in R^1. \tag{1.127}$$

The characteristic equation corresponding to equation (1.127) has the form

$$(i\sigma + i\lambda + a)(-\sigma^2 + \lambda^2 + ib\sigma - c^2) = 0, \quad \sigma \in R^1. \tag{1.128}$$

Let $\rho(\sigma)$ be the number of roots of the characteristic equation (1.128) with $Re\lambda \leq 0$ (for fixed $\sigma \in R^1$). It is easy to see that

$$\rho(\sigma) = \begin{cases} 2, & \text{for } \sigma > 0 \\ 1, & \text{for } \sigma < 0 \end{cases} \quad \text{and} \quad \rho(0) = \begin{cases} 3, & \text{for } c = 0 \\ 2, & \text{for } c \neq 0. \end{cases}$$

The boundary conditions (1.9) and (1.10) for (1.127) take the form

$$u(x,0) = f_0(x), \quad x \in R^1, \tag{1.129}$$

$$Re(b_1(x)\frac{\partial u(x,0)}{\partial t}) = f_1(x), \quad x \in R^1, \tag{1.130}$$

where $f_0(x)$, $f_1(x)$ and $b_1(x)$ are the same functions as in (1.9) and (1.10). By Theorem 1.1 the problem (1.127), (1.129), (1.130) is uniquely solvable in the class $M_0(R^2_+)$.

More examples of the correct boundary value problems for non-regular partial differential equations will be given in chapters 2 and 3.

Chapter 2

Riemann-Hilbert Problem for a Class of Non-Regular Elliptic Equations

2.1 Description of the Problem

Consider the following elliptic equation

$$\frac{\partial^n u}{\partial \bar{z}^n} + a_1 \frac{\partial^{n-1} u}{\partial \bar{z}^{n-1}} + \cdots + a_n u = 0, \quad x \in R^1, \quad y > 0, \tag{2.1}$$

where $a_1, \ldots, a_n$ are some real constants, $u(x, y)$ is a solution to be found and

$$\frac{\partial}{\partial \bar{z}} = \frac{1}{2}(\frac{\partial}{\partial x} + i\frac{\partial}{\partial y}).$$

For the boundary conditions of (2.1) we take

$$Re\left[b_k(x)\frac{\partial^k u(x,0)}{\partial y^k}\right] = f_k(x), \quad x \in R^1, k = 0, \ldots, n-1, \tag{2.2}$$

where $b_k(x)$ and $f_k(x)$ are similar to those in (1.10).

The problem (2.1), (2.2) is said to be homogeneous, if $f_k(x) \equiv 0$, $(k = 0, 1, \ldots, n-1)$.

In this chapter we prove the following

Theorem 2.1 *If the indices of functions $b_k(x)$ on X-axis are equal to zero $(k = 0, 1, \ldots, n-1)$ and all the roots of the equation*

$$\lambda^n + a_1\lambda^{n-1} + \cdots + a_n = 0 \tag{2.3}$$

are real, then the problem (2.1), (2.2) is uniquely solvable in the class $M_0(R^2_+)$.

At the end of this chapter we will also consider boundary value problems for equation (2.1) in the case, where the coefficients of equation (2.1) are arbitrary complex constants and some of the roots of equation (2.3) are not real.

2.2 The General Solution of Equation (2.1)

To investigate the problem (2.1), (2.2) we construct the general solution of equation (2.1) in the class $M_0(R^2_+)$ (cf. the definition of the classes $N_\alpha(R^1)$ and $M_\alpha(R^2_+)$ $(\alpha \geq 0)$ in section 1.1). Here R^2_+ denotes the half-plane $x \in R^1,\ y > 0$.

Let all the roots of equation (2.3) be real. First consider the equation

$$\frac{\partial u}{\partial \bar{z}} - \lambda u = 0, \quad x \in R^1, \quad y > 0, \tag{2.4}$$

where λ is a real constant.

Changing in (2.4) the variable

$$u = e^{-2i\lambda y} w, \tag{2.5}$$

we get

$$\frac{\partial w}{\partial \bar{z}} = 0,$$

so $w(x, y) = \varphi(z)$, where $\varphi(z)$ is an arbitrary analytic on $z = x + iy$ function. Hence the general solution of equation (2.4) is defined by

$$u(x, y) = e^{-2i\lambda y} \varphi(z). \tag{2.6}$$

If $u(x, y) \in M_0(R^2_+)$, then it follows from (2.6) that $\varphi(x + iy)$ also belongs to the same class.

Consider now the non-homogeneous equation

$$\frac{\partial u}{\partial \bar{z}} - \mu u = e^{-2i\lambda y} \varphi(z), \quad x \in R^1, \quad y > 0, \tag{2.7}$$

where $\varphi(z)$ is an analytic function from the class $M_0(R^2_+)$, while λ and μ $(\lambda \neq \mu)$ are real constants.

It is easy to check that a particular solution $u_0(x, y)$ of (2.7) is defined by

$$u_0(x, y) = \frac{1}{\lambda - \mu} e^{-2i\lambda y} \varphi(z). \tag{2.8}$$

Now we turn to the construction of the general solution of equation (2.1) in the half-plane $x \in R^1, y > 0$, when the roots $\lambda_1, \ldots, \lambda_n$ of equation (2.3) are real and simple.

Equation (2.1) can be written as follows

$$(\frac{\partial}{\partial \bar{z}} - \lambda_1 I) \cdots (\frac{\partial}{\partial \bar{z}} - \lambda_n I)u = 0, \tag{2.9}$$

where I is the identity operator.

Resolving the first-order equation and using (2.6) and (2.8), from (2.9) we find that the general solution of equation (2.1) is defined by

$$u(x, y) = \sum_{j=1}^{n} e^{-2i\lambda_j y} \varphi_j(z), \tag{2.10}$$

where $\varphi_1(z), \ldots, \varphi_n(z)$ are arbitrary analytic in R^2_+ functions.

If $u(x, y) \in M_0(R^2_+)$, then it follows from the method of construction of the general solution of (2.10) that all the analytic functions $\varphi_1(z), \ldots, \varphi_n(z)$ also belong to the same class.

To obtain the general solution of equation (2.1) in the case, where the roots of (2.3) are multiple, we consider the following first-order equation

$$\frac{\partial u}{\partial \bar{z}} - \mu u = y^m e^{-2i\lambda y} \varphi(z), \quad x \in R^1, \quad y > 0, \tag{2.11}$$

where $\varphi(z)$ is an analytic function from the class $M_0(R^2_+)$, λ and μ are real constants and m is a non-negative integer.

Let $\lambda = \mu$. It is easy to see that a particular solution of equation (2.11) is defined by

$$u(x, y) = -\frac{2i}{m+1} y^{m+1} e^{-2i\lambda y} \varphi(z). \tag{2.12}$$

Now let $\lambda \neq \mu$. A particular solution of equation (2.11) is searched as

$$u(x, y) = P_m(y) e^{-2i\lambda y} \varphi(z), \tag{2.13}$$

where $P_m(y)$ is a polynomial of degree m:

$$P_m(y) = c_0 + c_1 y + \cdots + c_m y^m. \tag{2.14}$$

Substituting $u(x, y)$ from (2.13) into equation (2.11) we get

$$\frac{1}{2} i(c_1 + 2c_2 y + \cdots + m c_m y^{m-1}) + (\lambda - \mu)(c_0 + c_1 y + \cdots + c_m y^m) = y^m. \tag{2.15}$$

Comparing the coefficients of y^k $(k = 0,\dots,m)$ on both sides of (2.15) we get a recurrent formula to determine the constants c_k $(k = 0,\dots,m)$:

$$c_m = \frac{1}{\lambda - \mu}, \tag{2.16}$$

$$c_k = \frac{i(k+1)}{2(\mu - \lambda)} c_{k+1}, \quad k = m-1, m-2, \dots, 0. \tag{2.17}$$

Hence, for $\lambda \neq \mu$ a particular solution of equation (2.11) is defined by (2.13), where the coefficients of the polynomial (2.14) are defined by (2.16) and (2.17).

Now let us consider the general case, where there are multiple roots of equation (2.3). Let $\lambda_1, \dots, \lambda_r$ be the roots of (2.3) ($\lambda_j \neq \lambda_k$ for $j \neq k$) with multiplicities $k_1, k_2, \dots, k_r$ respectively and $k_1 + k_2 + \dots + k_r = n$.

Arguing as in the case of simple roots and using the relations (2.12) and (2.13), we conclude that the general solution of equation (2.1) in the class $M_0(R^2_+)$ is defined by

$$u(x,y) = \sum_{j=1}^{r} \sum_{l=0}^{k_j - 1} e^{-2i\lambda_j y} y^l \varphi_{jl}(z), \tag{2.18}$$

where $\varphi_{jl}(z)$ are arbitrary analytic in R^2_+ functions belonging to $M_0(R^2_+)$.

2.3 Cauchy Problem for Equation (2.1)

To investigate the problem (2.1), (2.2) let us first consider the Cauchy problem for equation (2.1) in half-plane $x \in R^1$, $y > 0$. It is known that the boundary data for the Cauchy problem have the form

$$\frac{\partial^k u(x,0)}{\partial y^k} = \psi_k(x), \quad x \in R^1, \quad k = 0, 1, \dots, n-1, \tag{2.19}$$

where $\psi_k(x)$ are known functions defined on real axis.

We will assume that $\psi_k(x) \in N_0(R^1)$ and the solution $u(x,y)$ belongs to the class $M_0(R^2_+)$.

It follows from (2.18) that if $u(x,y)$ is a solution of equation (2.1) in R^2_+ and $u(x,y) \in M_0(R^2_+)$, then the functions

$$\frac{\partial^k u(x,0)}{\partial y^k} \quad (k = 0, 1, \dots, n-1)$$

are the boundary values of analytic in R^2_+ functions belonging to the class $M_0(R^2_+)$. So $\psi_k(x)$ in the conditions (2.19) are the boundary values of analytic in R^2_+ functions and $\psi_k(z) \in M_0(R^2_+)$.

One has the following

Theorem 2.2 *If all the roots of equation (2.3) are real, then the problem (2.1), (2.19) is uniquely solvable in the class* $M_0(R^2_+)$.

Proof. For simplicity, we prove the assertion of Theorem 2.2 in the case where all the roots $\lambda_1, \ldots, \lambda_n$ of equation (2.3) are simple. The general case can be treated in a similar way.

Let $u(x,y)$ be a solution of the problem (2.1), (2.19). We define the functions $w_0(x,y), \ldots, w_{n-1}(x,y)$ by

$$w_0(x,y) = u(x,y), \quad w_k(x,y) = (\frac{\partial}{\partial \bar{z}} - \lambda_1 I) \cdots (\frac{\partial}{\partial \bar{z}} - \lambda_k I) u(x,y), \tag{2.20}$$

where $k = 1, 2, \ldots, n-1$.

It is clear that

$$w_k(x,y) \equiv \frac{\partial^k u}{\partial \bar{z}^k} + a_{k1} \frac{\partial^{k-1} u}{\partial \bar{z}^{k-1}} + \cdots + a_{kk} u, \tag{2.21}$$

where $a_{k1}, a_{k2}, \ldots, a_{kk}$ are some real constants.

From (2.19) and (2.21) we have

$$w_k(x,y) = \omega_k(z) \quad \text{for} \quad y = 0, \quad x \in R^1, \quad k = 0, 1, \ldots, n-1, \tag{2.22}$$

where

$$\omega_0(z) = \psi_0(z), \quad \omega_k(z) = \psi_k(z) + a_{k1}\psi_{k-1}(z) + \cdots + a_{kk}\psi_0(z), \tag{2.23}$$

where $k = 1, \ldots, n-1$.

It is easy to see that conditions (2.19) and (2.22) are equivalent. In view of (2.20), equation (2.1) can be written as (cf. (2.9))

$$\frac{\partial w_{n-1}}{\partial \bar{z}} - \lambda_n w_{n-1} = 0. \tag{2.24}$$

The general solution of equation (2.24) is defined by

$$w_{n-1}(x,y) = e^{-2i\lambda_n y} \varphi_{n-1}(z), \tag{2.25}$$

where $\varphi_{n-1}(z)$ is an arbitrary analytic in R^2_+ function.

From the condition (2.22) with $k = n-1$, we get

$$\varphi_{n-1}(x) = \omega_{n-1}(x), \quad x \in R^1.$$

Since both functions $\varphi_{n-1}(z)$ and $\omega_{n-1}(z)$ are analytic, we have

$$\varphi_{n-1}(z) = \omega_{n-1}(z), \quad x \in R^1, \quad y > 0. \tag{2.26}$$

Using (2.20) the relation (2.25) can be written as follows:

$$\frac{\partial w_{n-2}}{\partial \bar{z}} - \lambda_{n-1} w_{n-2} = e^{-2i\lambda_n y} \varphi_{n-1}(z). \tag{2.27}$$

It follows from (2.6) and (2.8) that the general solution of equation (2.27) is defined by

$$w_{n-2}(x,y) = \frac{1}{\lambda_n - \lambda_{n-1}} e^{-2i\lambda_n y} \omega_{n-1}(z) + e^{-2i\lambda_{n-1} y} \varphi_{n-2}(z), \tag{2.28}$$

where $\varphi_{n-2}(z)$ is an arbitrary analytic in R_+^2 function.

From (2.22) with $k = n-2$ and (2.28) we get

$$\varphi_{n-2}(z) = \omega_{n-2}(z) - \frac{1}{\lambda_n - \lambda_{n-1}} \omega_{n-1}(z). \tag{2.29}$$

Since $\psi_k(z) \in M_0(R_+^2)$ $(k = 0, \ldots, n-1)$, from (2.23), (2.26) and (2.29) it follows that $\varphi_k(z) \in M_0(R_+^2)$ for $k = n-1, n-2$.

Continuing in the same manner we get the desired assertion.

Theorem 2.2 is proved.

2.4 Riemann-Hilbert Problem for Analytic Functions

Consider the following problem

Problem A. Find an analytic in the half-plane $Imz > 0$ function $\varphi(z)$ belonging to the class $M_0(R_+^2)$ and satisfying the boundary condition

$$Re\,[b(x)\varphi(x)] = f(x), \quad x \in R^1, \tag{2.30}$$

where $b(x)$ and $f(x)$ are given functions, $f(x)$ being real-valued. We suppose that the functions $b(x)$ and $f(x)$ satisfy the same conditions as $b_k(x)$ and $f_k(x)$ in section 2.1 (see condition (2.2)).

Problem A in the class of bounded function was exhaustively investigated in the monograph [15]. This problem in the class $M_0(R_+^2)$ can be solved in a similar way. For the sake of completeness, we are going to sketch a solution of this problem, omitting the details.

According to the assumption, $b(x) \neq 0$ for $x \in R^1$ and the index of $b(x)$ on X-axis is equal to zero,

$$\lim_{|x|\to\infty} b(x) = b(\infty) \neq 0$$

and $b(x) - b(\infty) \in N_0(R^1)$. Thus, the function

$$\theta(x) = arg\frac{b(x)}{b(\infty)} \tag{2.31}$$

belongs to the class $N_0(R)$. Consider the functions

$$\omega(z) = \frac{1}{\pi i}\int_{-\infty}^{+\infty}\frac{\theta(\zeta)d\zeta}{\zeta - z}, \quad \Omega(z) = exp(i\omega(z)). \tag{2.32}$$

Since $\theta(\zeta) \in N_0(R^1)$ the function $\omega(z)$ satisfies the estimates (see [15])

$$|\, \omega^{(k)}(z)\, | \leq c_k(1+ |\, z\, |)^{\alpha_k}, \quad k = 0, 1, \ldots, \tag{2.33}$$

where c_k and α_k are some negative constants depending on $b(x)$.

Denote by $\omega(x)$ and $\Omega(x)$ the boundary values of $\omega(z)$ and $\Omega(z)$ respectively, as $z \to x$ and $Imz > 0$. It is clear that $\Omega(z) \neq 0$ for $Imz \geq 0$ and $\Omega(x) = exp(i\omega(x))$. From (2.33) it follows that $\Omega(z) \to 1$ as $|\, z\, | \to +\infty$, $Imz \geq 0$ and the functions $\Omega(z) - 1$ and $[\Omega(z)]^{-1} - 1$ and their derivatives satisfy the estimates of the form (2.33).

Since $\theta(\zeta)$ is a real-valued function, then (see [15])

$$Re\omega(x) = \theta(x), \quad x \in R^1. \tag{2.34}$$

By assumption $|\, b(x)\, | = 1$. Thus

$$\frac{b(x)}{b(\infty)} = e^{i\theta(x)}. \tag{2.35}$$

It follows from (2.34) that

$$\theta(x) = \omega(x) - i\beta(x), \tag{2.36}$$

where $\beta(x) = Im\omega(x)$. Substituting $\theta(x)$ from (2.36) into (2.35) we get

$$b(x) = \Omega(x)e^{\beta(x)}b(\infty). \tag{2.37}$$

By (2.37) the condition (2.30) can be represented as:

$$Re\varphi_0(x) = f_0(x), \quad x \in R^1, \tag{2.38}$$

where

$$\varphi_0(z) = b(\infty)\Omega(z)\varphi(z), \tag{2.39}$$

$$f_0(x) = f(x)e^{-\beta(x)}. \tag{2.40}$$

Since $\varphi(z) \in M_0(R^2_+)$ and $f \in N_0(R^1)$, then from (2.39) and (2.40) it follows that $\varphi_0(z) \in M_0(R^2_+)$ and $f_0 \in N_0(R^1)$.

A particular solution of the problem (2.38) is defined by (see [14])

$$\varphi_0(z) = \frac{1}{\pi i} \int_{-\infty}^{+\infty} \frac{f_0(\zeta)d\zeta}{\zeta - z}. \tag{2.41}$$

According to (2.33), this solution belongs to the class $N_0(R^2_+)$.

Now we find the general solution of the problem (2.38) for $f_0 \equiv 0$. Since $\varphi_0(z) \in M_0(R^2_+)$, then

$$|\varphi_0(z)| \leq c(1+|z|)^{m-1},$$

where m is an integer. So a solution of the homogeneous problem (2.38) can be represented as (see [14]):

$$\varphi_0(z) = ic_0 + ic_1 z + \cdots + ic_{m-1} z^{m-1},$$

where $c_0, \ldots, c_{m-1}$ are some real constants.

From here, in particular, we get

$$\varphi_0(x) = ic_0 + ic_1 x + \cdots + ic_{m-1} x^{m-1}. \tag{2.42}$$

Since $\varphi_0(z) \in M_0(R^2_+)$, we have $\varphi_0(x) \to 0$ as $|x| \to +\infty$. From here and (2.42) it follows that $c_0 = c_1 = \cdots = c_{m-1} = 0$.

Thus, the homogeneous problem (2.38) in this class has only zero solution. This means that the problem (2.30) has a unique solution which is defined by

$$\varphi(z) = [b(\infty)\Omega(z)]^{-1} \frac{1}{\pi i} \int_{-\infty}^{+\infty} \frac{f_0(\zeta)d\zeta}{\zeta - z}, \quad Imz > 0. \tag{2.43}$$

In particular, it follows from (2.43) that the solution $\varphi(z)$ and its derivatives satisfy estimates of the form (2.33) and $\varphi(z) \in N_0(R^2_+)$.

Remark 2.1. If $\varphi(z)$ is analytic in the half-plane $x \in R$, $y > 0$ and $\varphi(z) \in M_0(R^2_+)$, then $\varphi(z) \in N_0(R^2_+)$.

Indeed, by the definition of the class $M_0(R^2_+)$ we have $Re\varphi(x) = f_0(x)$, where $f(x) \in N_0(R^1)$. Hence $\varphi(z)$ is determined by $f_0(x)$ via formula (2.43), where $b(\infty) = 1$ and $\Omega(z) = 1$. From (2.43) it immediately follows that $\varphi(x) \in N_0(R^2_+)$.

2.5 Investigation of the Problem (2.1), (2.2)

Let $\varphi_k(z)$ be a solution of the problem A (cf. section 2.4) for $b(x) = b_k(x)$ and $f(x) = f_k(x)$ $(k = 0, 1, \ldots, n-1)$. Let the right-hand sides of the boundary conditions (2.19) coincide with these functions.

One has the following

Theorem 2.3 *The problem (2.1), (2.2) is equivalent to the Cauchy problem (2.1), (2.19).*

Proof. Let $u(x, y)$ be a solution of the problem (2.1), (2.19) in the class $M_0(R^2_+)$. Substituting the values of

$$\frac{\partial^k u(x,0)}{\partial y^k}, \quad k = 0, 1, \ldots, n-1$$

from (2.19) into (2.2) and taking into account that $\psi_k(z)$ is a solution of problem A for $b(x) = b_k(x)$ and $f(x) = f_k(x)$, we see that $u(x, y)$ satisfies the boundary conditions (2.2). So a solution of the Cauchy problem (2.1), (2.19) is a solution of the problem (2.1), (2.2).

Now we prove the inverse assertion. Let $u(x, t)$ be a solution of the problem (2.1), (2.2) belonging to the class $M_0(R^2_+)$. By (2.18) we have

$$\frac{\partial^k u(x,0)}{\partial y^k} = \psi_k(x) \quad \text{for} \quad y = 0, \quad k = 0, \cdots, n-1, \tag{2.44}$$

where $\psi_0(z), \psi_1(z), \cdots, \psi_{n-1}(z)$ are some analytic functions from the class $M_0(R^2_+)$. Substituting the values of

$$\frac{\partial^k u(x,0)}{\partial y^k}, \quad k = 0, 1, \ldots, n-1$$

from (2.44) into the condition (2.2) we see that $\psi_k(z)$ is a solution of problem A for $b(x) = b_k(x)$ and $f(x) = f_k(x)$. So $u(x, t)$ is a solution of the Cauchy problem (2.1), (2.19).

Theorem 2.3 is proved.

Thus, the problem (2.1), (2.2) can be reduced to the Cauchy problem (2.1), (2.19) which is uniquely solvable in the class $M_0(R^2_+)$.

Thus we have proved Theorem 2.1.

2.6 General Boundary Value Problems for Equation (2.1)

In this section we take the following boundary conditions for equation (2.1),

$$\sum_{j=0}^{k} Re\left[b_{kj}(x)\frac{\partial^j u(x,0)}{\partial y^k}\right] = f_k(x), \quad x \in R^1, \quad k = 0,\cdots,n-1, \tag{2.45}$$

where $b_{kk}(x)$ and $f_k(x)$ $(k = 0,1,\cdots,n-1)$ satisfy the same conditions as the functions $b_k(x)$ and $f_k(x)$ in (2.2). For $0 \le j \le k-1$ we impose on $b_{kj}(x)$ the restrictions

$$\left|\frac{d^k b_{kj}(x)}{dx^k}\right| \le c_k, \quad k = 0,1,\ldots,$$

where c_k are some constants.

One has the following

Theorem 2.4 *The problem (2.1), (2.45) is uniquely solvable in the class $M_0(R^2_+)$.*

To prove Theorem 2.4 and to construct a solution, we consider the following problem.

Problem B. Find analytic in R^2_+ functions $\psi_0(z), \psi_1(z), \cdots, \psi_{n-1}(z)$, belonging to the class $M_0(R^2_+)$ and satisfying the boundary conditions

$$\sum_{j=0}^{k} Re[b_{kj}(x)\psi_j(x)] = f_k(x), \quad x \in R^1, \quad k = 0,\cdots,n-1. \tag{2.46}$$

Problem B has a solution and it is unique. This assertion immediately follows from the existence and uniqueness of a solution of problem A (cf. section 2.4).

Indeed, for $k = 0$ from (2.46) we determine a uniquely analytic function $\psi_0(z)$. Substituting $\psi_0(z)$ into (2.46) at $k = 1$ we find $\varphi_r(z)$, etc.

Next, let the functions $\psi_0(z),\cdots,\psi_{n-1}(z)$ on the right-hand side of the condition (2.19) satisfy problem B. Then one has the following

Theorem 2.5 *The problem (2.1), (2.45) is equivalent to the problem (2.1), (2.19).*

Proof is similar to that of Theorem 2.3, and so is omitted.

Thus, the resolution of the problem (2.1), (2.45) is reduced to that of Cauchy problem (2.1), (2.19), considered in section 2.3. From here and from the existence and uniqueness of a solution of Cauchy problem (2.1), (2.19), Theorem 2.4 follows.

2.7 Exceptional Cases for Problem (2.1), (2.2)

Assume that some of the roots of equation (2.3) are complex numbers. Will the problem (2.1), (2.2) be correct in this case?

Below we will show that this essentially depends on the coefficients of equation (2.1). To this end we consider the following problems.

Problem 1.

$$\frac{\partial u}{\partial \bar{z}} + a_1 u = 0, \quad x \in R^1, \quad y > 0, \tag{2.47}$$

$$Reu(x,0) = f_0(x), \quad x \in R^1, \tag{2.48}$$

where a_1 is a constant such that $Ima_1 \neq 0$ and $f_0(x) \in N_0(R^1)$.

Problem 2.

$$\frac{\partial^2 u}{\partial \bar{z}^2} + b_1 \frac{\partial u}{\partial \bar{z}} + b_2 u = 0, \quad x \in R^1, \quad y > 0, \tag{2.49}$$

$$Reu(x,0) = f_0(x), \tag{2.50}$$

$$Re\left[i\frac{\partial u(x,0)}{\partial y} + iku(x,0)\right] = f_1(x), \tag{2.51}$$

where b_1, b_2 and k are real constants, $f_j(x) \in N_0(R^1)$ $(j = 1, 2)$ and $b_1^2 < 4b_2$.

The solutions of problems 1 and 2 are searched in the class $M_0(R^2_+)$.

It is clear that the roots of the equations $\lambda + a_1 = 0$ and $\lambda^2 + b_1\lambda + b_2 = 0$ are not real.

The aim of this section is to show that problem 1 is incorrect, while the correctness of problem 2 depends on constant k.

Investigation of the problem 1. Case I. Assume that $Ima_1 > 0$. Let us consider the following analytic function

$$\varphi(z) = \int_{-\infty}^{+\infty} \psi(t)e^{izt}\,dt,$$

where $\psi(t)$ is an arbitrary infinitely differentiable real-valued odd function such that $\psi(t) = 0$ for $|\, t \,| \geq Ima_1$. It is clear that $Re\varphi(x) = 0$ for $x \in R^1$, and

$$u(x,y) = e^{2a_1 iy}\varphi(z)$$

is a solution of the homogeneous problem 1 (at $f_0 \equiv 0$). So the homogeneous problem 1 for $Ima_1 > 0$ has infinitely many linearly independent solutions.

Case II. Let $Ima_1 < 0$. We are going to show that for finite function $f_0(x)$, $f_0(x) \neq 0$, the non-homogeneous problem (2.47), (2.48) has no solution.

The finiteness of a function $f_0(x)$ means that $f_0(x) = 0$ for $|x| \geq x_0$, where x_0 is a positive constant.

Let $u(x, y)$ be a solution of the problem (2.47), (2.48). According to (2.6) this solution can be represented as

$$u(x,y) = e^{2a_1 iy}\varphi(z), \tag{2.52}$$

where $\varphi(z)$ is analytic in R_2^+ function.

From (2.52) we have

$$\varphi(z) = u(x,y)e^{-2a_1 iy}. \tag{2.53}$$

Since $u(x, y)$ belongs to the class $M_0(R_+^2)$, it follows from (2.53) that $\varphi(z)$ also belongs to the same class.

From (2.48) and (2.52) we have

$$Re\varphi(x) = f_0(x), \quad x \in R^1. \tag{2.54}$$

Hence (see [14])

$$\varphi(z) = \frac{1}{\pi i}\int_{-x_0}^{+x_0} \frac{f_0(t)dt}{t-z}, \quad Imz > 0. \tag{2.55}$$

From (2.55) it follows that $\varphi(z)$ admits Laurent expansion at the vicinity of infinity

$$\varphi(z) = \sum_{k=1}^{\infty} \frac{c_k}{z^k}, \quad |z| > x_0, \tag{2.56}$$

where c_k are some constants and the series converges uniformly with respect to z for $|z| \geq 2x_0$. Since $u(x,y) \in M_0(R_+^2)$, then from (2.53) we have

$$|\varphi(iy)| \leq c(1+|y|)^m \exp(2Ima_1 y), \tag{2.57}$$

where c and m are some positive constants.

Substituting $\varphi(z)$ from (2.56) into (2.57) we get

$$\left|\sum_{k=1}^{\infty} \frac{c_k}{(iy)^k}\right| \leq c(1+|y|)^m \exp(2Ima_1 y) \quad for \quad y > x_0. \tag{2.58}$$

Multiplying both sides of (2.58) by y^k $(k = 1, 2, \ldots)$ and passing to the limit as $y \to +\infty$ we get $c_k = 0$, $k = 1, 2, \ldots$. We have $\varphi(z) \equiv 0$ and $u(x, y) \equiv 0$. So the boundary condition (2.48) implies $f_0(x) \equiv 0$. Hence in this case the non-homogeneous problem (2.47), (2.48) has no solution for every finite infinitely differentiable function $f_0(x)$, $(f_0(x) \neq 0)$.

Thus, if $Im a_1 \neq 0$ the problem 1 is incorrect. In a similar way we can prove that this problem is also incorrect in the case, where the boundary conditions have the form

$$Re[b(x)u(x,0)] = f_0(x), \quad x \in R^1,$$

where the function $b(x)$ satisfies the same conditions as the function $b_k(x)$ in conditions (2.2).

Investigation of problem 2. Let $\lambda_1 = \alpha + i\beta$ and $\lambda_2 = \alpha - i\beta$ $(\beta > 0)$ be the roots of the equation $\lambda^2 + b_1\lambda + b_2 = 0$.

According to (2.10), the general solution of equation (2.49) is defined by

$$u(x,y) = e^{-2i\lambda_1 y}\varphi_1(z) + e^{-2i\lambda_2 y}\varphi_2(z), \tag{2.59}$$

where $\varphi_1(z)$ and $\varphi_2(z)$ are arbitrary analytic in the half-plane $x \in R^1$, $y > 0$ functions.

Rewrite (2.59) as follows

$$u(x,y) = e^{2i(\beta x - \alpha y)}\psi_1(z) + e^{-2i(\beta x + \alpha y)}\psi_2(z), \tag{2.60}$$

where

$$\psi_1(z) = \varphi_1(z)e^{-2i\beta z} \quad \text{and} \quad \psi_2(z) = \varphi_2(z)e^{2i\beta z}. \tag{2.61}$$

So the general solution of equation (2.49) in R^2_+ is defined by (2.60) where $\psi_1(z)$ and $\psi_2(z)$ are arbitrary analytic functions.

From (2.60) we get

$$\frac{\partial u(x,y)}{\partial \bar{z}} - \lambda_1 u = -i\beta e^{-2i(\beta x + \alpha y)}\psi_2(z), \tag{2.62}$$

$$\frac{\partial u(x,y)}{\partial \bar{z}} - \lambda_2 u = i\beta e^{2i(\beta x - \alpha y)}\psi_1(z). \tag{2.63}$$

From (2.62), (2.63) it immediately follows that if $u(x,y) \in M_0(R^2_+)$, then $\psi_1(z)$ and $\psi_2(z)$ also belong to the same class.

Substituting the general solution (2.60) into the boundary conditions (2.50), (2.51) we get

$$Re[e^{2i\beta x}\psi_1(x) + e^{-2i\beta x}\psi_2(x)] = f_0(x), \qquad x \in R^1, \tag{2.64}$$

$$\begin{aligned} &Re[(2\alpha + ik)(e^{2i\beta x}\psi_1(x) + e^{-2i\beta x}\psi_2(x))] \\ &-Re[e^{2i\beta x}\psi_1'(x) + e^{-2i\beta x}\psi_2'(x)] = f_1(x), \quad x \in R^1. \end{aligned} \tag{2.65}$$

Now, multiplying both sides of (2.64) by 2α and subtracting (2.65) from (2.64) we obtain

$$\begin{aligned} Re[e^{2i\beta x}\psi_1'(x) + e^{-2i\beta x}\psi_2'(x) - ik(e^{2i\beta x}\psi_1(x) + e^{-2i\beta x}\psi_2(x))] \\ = 2\alpha f_0(x) - f_1(x), \qquad x \in R^1. \end{aligned} \tag{2.66}$$

Differentiating both sides of (2.64) with respect to x and subtracting (2.66) from (2.64) we get

$$Re[i(k+2\beta)e^{2i\beta x}\psi_1(x)+i(k-2\beta)e^{-2i\beta x}\psi_2(x)] = f_0'(x)-2\alpha f_0(x)+f_1(x), \qquad x\in R^1. \tag{2.67}$$

It is clear that the conditions (2.64), (2.65) are equivalent to the conditions (2.64), (2.67).

We consider two cases.

Case I. Let $k\neq 2\beta$. Then dividing both sides of (2.67) by $2\beta-k$ we get

$$Re[i\gamma e^{2i\beta x}\psi_1(x)-ie^{-2i\beta x}\psi_2(x)] = g_1(x), \tag{2.68}$$

where

$$\gamma = \frac{2\beta+k}{2\beta-k} \quad \text{and} \quad g_1(x) = \frac{1}{-k+2\beta}[f_0'(x)-2\alpha f_0(x)+f_1(x)]. \tag{2.69}$$

Since $\beta>0$ we have $\gamma\neq -1$. Introduce a new function

$$\Phi_1(z) = \psi_2(z)+\mu e^{4i\beta z}\psi_1(z), \tag{2.70}$$

where μ is a real constant which will be chosen below.

Since $\psi_1(z)$ and $\psi_2(z)$ belong to the class $M_0(R^2_+)$ and $\beta>0$, then $\Phi_1(z)\in M_0(R^2_+)$. From (2.70) we have

$$\psi_2(z) = \Phi_1(z)-\mu e^{4i\beta z}\psi_1 z. \tag{2.71}$$

Substituting $\psi_2(z)$ from (2.71) into (2.64) and (2.68) we get

$$Re[e^{2i\beta x}\psi_1(x)(1-\mu)+e^{-2i\beta x}\Phi_1(x)] = f_0(x), \quad x\in R^1, \tag{2.72}$$

$$Re[(\gamma+\mu)ie^{2i\beta x}\psi_1(x)-ie^{-2i\beta x}\Phi_1(x)] = g_1(x), \quad x\in R^1. \tag{2.73}$$

We choose the constant μ to satisfy

$$1-\mu = \gamma+\mu \tag{2.74}$$

that is, we have

$$\mu = \frac{1-\gamma}{2} \quad \text{and} \quad 1-\mu = \gamma+\mu = \frac{\gamma+1}{2}\neq 0. \tag{2.75}$$

For this choice of μ conditions (2.72) and (2.73) are equivalent to condition:

$$(\frac{\gamma+1}{2}\psi_1(x)+\overline{\phi_1(x)})e^{2i\beta x} = f_0(x)-ig_1(x). \tag{2.76}$$

The condition (2.76) can be written as follows

$$\frac{\gamma+1}{2}\psi_1(x) + \overline{\phi_1(x)} = g(x), \quad x \in R^1, \tag{2.77}$$

where

$$g(x) = e^{-2i\beta x}(f_0(x) - ig_1(x)). \tag{2.78}$$

Denote

$$\Phi_2(z) = -\overline{\Phi_1(\overline{z})}, \quad Imz < 0. \tag{2.79}$$

Since $\phi_1(z)$ is analytic in $Imz > 0$ and belongs to the class $M_0(R^2_+)$, then $\phi_2(z)$ is analytic in $Imz < 0$ and $\phi_2(z) \in M_0(R^2_-)$, where $R^2_- = \{z : \ Imz < 0\}$. Therefore $\psi_1(z) \in N_0(R^2_+)$ and $\phi_2(z) \in N_0(R^2_-)$ (cf. Remark 2.1).

In view of (2.79) condition (2.77) can be written as

$$\frac{\gamma+1}{2}\psi_1(x) - \Phi_2(x) = g(x), \quad x \in R^1. \tag{2.80}$$

It is known (see [15]) that a solution of the conjugate problem (2.80) is defined by

$$\psi_1(z) = \frac{1}{(\gamma+1)\pi i}\int\limits_{-\infty}^{+\infty}\frac{g(\zeta)d\zeta}{\zeta - z}, \quad Imz > 0, \tag{2.81}$$

$$\Phi_2(z) = \frac{1}{2\pi i}\int\limits_{-\infty}^{+\infty}\frac{g(\zeta)d\zeta}{\zeta - z}, \quad Imz < 0. \tag{2.82}$$

From (2.79), (2.81) and (2.82) we get

$$\psi_1(z) = \frac{1}{\pi i(\gamma+1)}\int\limits_{-\infty}^{+\infty}\frac{g(\zeta)d\zeta}{\zeta - z}, \quad Imz > 0, \tag{2.83}$$

$$\Phi_1(z) = \frac{1}{2\pi i}\int\limits_{-\infty}^{+\infty}\frac{\overline{g(\zeta)}d\zeta}{\zeta - z}, \quad Imz > 0. \tag{2.84}$$

Substituting the values of $\psi_1(z)$ and $\Phi_1(z)$ from (2.83) and (2.84) into (2.71) we get

$$\psi_2(z) = \frac{1}{2\pi i}\int\limits_{-\infty}^{+\infty}\frac{\overline{g(\zeta)}d\zeta}{\zeta - z} - \frac{\mu\exp(4i\beta z)}{\pi(\gamma+1)i}\int\limits_{-\infty}^{+\infty}\frac{g(\zeta)d\zeta}{\zeta - z}, \quad Imz > 0. \tag{2.85}$$

From (2.69) and (2.78) we obtain

$$g(x) = e^{-2i\beta x}\left[f_0(x) - \frac{i}{2\beta - k}(f_0'(x) - 2\alpha f_0(x) + f_1(x))\right], \tag{2.86}$$

Thus, we have proved the following result.

Theorem 2.6 *If $k \neq \sqrt{4b_2 - b_1^2}$, then problem 2 is uniquely solvable and the solution is defined by (2.60), where the functions $\psi_1(z)$, $\psi_2(z)$, $g(x)$ and the constant γ are defined by (2.83), (2.85), (2.86) and (2.69) respectively.*

Case II. Let $k = 2\beta$. We prove that in this case the homogeneous problem 2 ($f_0 \equiv 0$ and $f_1(x) \equiv 0$) has infinitely many linearly independent solutions.

To this end we consider the following problem

$$Re[e^{-2i\beta x}\psi(x)] = 0, \quad x \in R^1, \tag{2.87}$$

where $\psi(z)$ is unknown analytic function and $\psi(z) \in N_0(R^2_+)$.

Consider an analytic function

$$\psi(z) = e^{2i\beta z}\int_{-1}^{+1}\omega(\tau)e^{-2i\beta\tau z}\,d\tau, \tag{2.88}$$

where $\omega(\tau)$ is an arbitrary infinitely differentiable real-valued odd function in R^1 and $\omega(\tau) \equiv 0$ for $|\tau| \geq 1$.

From the conditions imposed on $\omega(\tau)$ it follows that $\psi(z) \in N_0(R^2_+)$ and the condition (2.87) is fulfilled.

If $k = 2\beta$ and $\psi(z)$ is a solution of the boundary value problem (2.87), then it is easy to see that $\psi_1(z) = 0$ and $\psi_2(z) = \psi(z)$ satisfy the conditions (2.64) and (2.67) for $f_0 = f_1 = 0$. So each function of the form

$$u(x, y) = \psi(z)e^{-2i(\beta x + \alpha y)}, \tag{2.89}$$

is a solution of the homogeneous problem 2, where $\psi(z)$ is defined by (2.88). Hence for $k = 2\beta$ the homogeneous problem 2 has infinitely many linearly independent solutions.

Thus, we can mention a correct boundary value problem of the form (2.1), (2.45) in the case where the coefficients of equation (2.1) are real. We do not assume here that all the roots of equation (2.3) are real.

Let $\lambda_1, \ldots, \lambda_r$ be real and $\mu_1, \ldots, \mu_p$ and $\overline{\mu}_1, \ldots, \overline{\mu}_p$ be the complex roots of equation (2.3):

$$\mu_j = \alpha_j + i\beta_j, \quad \overline{\mu}_j = \alpha_j - i\beta_j, \quad j = 1, \ldots, p, \tag{2.90}$$

where $\alpha_1, \ldots, \alpha_p$, $\beta_1, \ldots, \beta_p$ are real numbers, $\beta_j > 0$, $j = 1, \ldots, p$ and $r + 2p = n$. The roots are counted according to their multiplicities.

Denote by L_j the operator

$$L_j u \equiv (\frac{\partial}{\partial \bar{z}} - \mu_j I)(\frac{\partial}{\partial \bar{z}} - \overline{\mu}_j I)u, \quad j = 1, \ldots, p, \tag{2.91}$$

where I is the identity operator.

Since $\mu_j = \alpha_j + i\beta_j$ and $\overline{\mu}_j = \alpha_j - i\beta_j$, the operator L_j can be represented as follows

$$L_j u \equiv \frac{\partial^2 u}{\partial \bar{z}^2} - 2\alpha_j \frac{\partial u}{\partial \bar{z}} + (\alpha_j^2 + \beta_j^2)u, \quad j = 1, \ldots, p. \tag{2.92}$$

Let

$$(\lambda - \lambda_1) \cdots (\lambda - \lambda_r) \equiv \lambda^r + a_{r1}\lambda^{r-1} + \cdots + a_{rr}, \tag{2.93}$$

where $a_{r1}, \ldots, a_{rr}$ are some real constants.

Denote by L_0 the operator

$$L_0 u \equiv \frac{\partial^r u}{\partial \bar{z}^r} + a_{r1} \frac{\partial^{r-1} u}{\partial \bar{z}^{r-1}} + \cdots + a_{rr} u. \tag{2.94}$$

Consider the following boundary conditions

$$Re \left[b_k(x) \frac{\partial^k u(x,0)}{\partial y^k} \right] = f_k(x), \quad x \in R^1, \quad k = 0, \ldots, r-1, \tag{2.95}$$

$$Re[L_\nu L_{\nu-1} \cdots L_0 u(x,y)] = g_{1\nu}(x) \quad \text{for} \quad y = 0, \tag{2.96}$$

$$Re \left[i \frac{\partial}{\partial y} L_\nu L_{\nu-1} \cdots L_0 u(x,y) \right] = g_{2\nu}(x) \quad \text{for } y = 0, x \in R^1, \tag{2.97}$$

where $f_k(x)$ $(k = 0, \ldots, r-1)$, $g_{1\nu}(x)$ and $g_{2\nu}(x)$ $(\nu = 0, \ldots, p-1)$ belong to the class $N_0(R^1)$, and $b_k(x)$ $(k = 0, \ldots, r-1)$ satisfy the same conditions as in (2.2).

A solution of the problem (2.1), (2.95) – (2.97) is searched in the class $M_0(R_+^2)$. We also assume that the index of function $f_k(x)$ on X-axis is equal to zero. It is clear that the boundary conditions (2.95) – (2.97) have the form (2.45).

One has the following

Theorem 2.7 *If the coefficients of equation (2.1) are real, then the problem (2.1), (2.95) – (2.97) is uniquely solvable.*

Proof. Denote

$$w_1 = L_{p-1} L_{p-2} \cdots L_0 u. \tag{2.98}$$

Equation (2.1) can be written as follows:

$$L_p w_1 = 0. \tag{2.99}$$

The boundary conditions (2.96), (2.97) for $\nu = p-1$ take the form

$$Re w_1(x,0) = g_{1,p-1}(x), \tag{2.100}$$

$$Re\left[\frac{i\partial w_1(x,0)}{\partial y}\right] = g_{2,p-1}(x). \tag{2.101}$$

Since the solution $u(x,y)$ belongs to the class $M_0(R^2_+)$ then $w_1(x,y)$ also belongs to the same class.

Hence problem 2 can be used to define $w_1(x,y)$. As we have shown above, problem 2 is uniquely solvable and its solution has the form (see (2.60)):

$$w_1(x,y) = \psi_{11}(z)\exp[2i(\beta_p x - \alpha_p y)] + \psi_{12}(x)\exp[-2i(\beta_p x + \alpha_p y)], \tag{2.102}$$

where $\psi_{11}(z)$ and $\psi_{12}(z)$ are some analytic in R^2_+ functions belonging to the class $M_0(R^2_+)$. So the left-hand side of (2.98) is known and is defined by (2.102).

Denote

$$w_2(x,y) \equiv L_{p-2}\cdots L_0 u. \tag{2.103}$$

Equation (2.98) takes the form

$$L_{p-1}w_2(x,y) = w_1(x,y), \tag{2.104}$$

where $w_1(x,y)$ is defined by (2.102).

Conditions (2.96), (2.97) for $\nu = p-2$ can be represented as

$$Re w_2(x,0) = g_{1,p-2}(x), \tag{2.105}$$

$$Re\left[i\frac{\partial w_2(x,0)}{\partial y}\right] = g_{2,p-2}(x). \tag{2.106}$$

Thus, we obtain the problem (2.104), (2.105), (2.106) to determine $w_2(x,y)$. Equation (2.104) with respect to $w_2(x,y)$ can be rewritten as follows:

$$(\frac{\partial}{\partial \bar{z}} - \overline{\mu}_{p-1}I)(\frac{\partial}{\partial \bar{z}} - \mu_{p-1}I)w_2(x,y) = w_1(x,y). \tag{2.107}$$

Later we will find a particular solution of equation (2.114) in the class $M_0(R^2_+)$ (cf. (2.115)). Similarly we find a particular solution of equation (2.107) in the same class.

Let $w_0(x,y)$ be a particular solution of equation (2.107) in the class $M_0(R^2_+)$. Setting

$$w_2(x,y) = w(x,y) + w_0(x,y) \tag{2.108}$$

into the problem (2.104) – (2.106) we get

$$L_{p-1}w(x,y) = 0, \quad x \in R^1, \quad y > 0, \tag{2.109}$$
$$Rew(x,0) = \mu_1(x), \quad x \in R^1, \tag{2.110}$$
$$Rei\frac{\partial w(x,0)}{\partial y} = \mu_2(x), \quad x \in R^1, \tag{2.111}$$

where

$$\mu_1(x) = g_{1,p-2}(x) - Re[w_0(x,0)], \tag{2.112}$$
$$\mu_2(x) = g_{2,p-2}(x) - Re\left[i\frac{\partial w_0(x,0)}{\partial y}\right]. \tag{2.113}$$

By Theorem 2.6 the problem (2.109) – (2.111) in the class $M_0(R^2_+)$ can be solved uniquely.
Substituting $w_2(x,y)$ from (2.108) into (2.103) we get

$$L_{p-2}\cdots L_0 u = w(x,y) + w_0(x,y).$$

Continuing the same process we get a solution of the problem (2.1), (2.95) – (2.97) which is defined uniquely.
Theorem 2.7 is proved.

2.8 Correct Boundary Value Problems for Equation (2.1) in General

Now let the coefficients of equation (2.1) be arbitrary complex numbers. First, consider the non-homogeneous first-order equation

$$\frac{\partial u}{\partial \bar{z}} - \lambda u = \sum_{k=1}^{m} P_k(y) e^{2i(\beta_k x - \alpha_k y)} \psi_k(z), \tag{2.114}$$

where $\lambda = \alpha + i\beta$, $p_1(y), \ldots, p_m(y)$ are given polynomials, $\alpha_1, \ldots, \alpha_m$, $\beta_1, \ldots, \beta_n$ are real constants and $\psi_1(z), \ldots, \psi_m(z)$ are given analytic in R^2_+ functions belonging to the class $N_0(R^2_+)$.

Lemma 2.1 *Equation (2.114) has a particular solution $u_0(x,y)$ of the form*

$$u_0(x,y) = \sum_{k=1}^{m} Q_k(y) e^{2i(\beta_k x - \alpha_k y)} \psi_k(z), \tag{2.115}$$

where $Q_1(y), \ldots, Q_m(y)$ are some polynomials and if $\lambda \neq \alpha_k + i\beta_k$, the degree of polynomial $Q_k(y)$ coincides with the degree of $P_k(y)$ and if $\lambda = \alpha_k + i\beta_k$, the degree of polynomial $Q_k(y)$ is superior to that of $P_k(y)$ by one.

Proof. Equation (2.114) can be written as

$$\frac{\partial u}{\partial \bar{z}} - \lambda u = \sum_{k=1}^{m} P_k(y) e^{-2i\lambda_k y} \varphi_k(z), \tag{2.116}$$

where $\lambda_k = \alpha_k + i\beta_k$ and

$$\varphi_k(z) = \psi_k(z) e^{2i\beta_k z}. \tag{2.117}$$

In section 2.2 we have constructed a particular solution of equation (2.11) (cf. (2.12) and (2.13)) implying that equation (2.114) has a solution of the form (2.115).

Lemma 2.1 is proved.

We take now the boundary condition for non-homogeneous equation (2.114) as

$$Re[e^{-2i\beta x} b(x) u(x,0)] = f(x), \quad x \in R^1, \tag{2.118}$$

where $b(x)$ and $f(x)$ satisfy the same conditions as $b_k(x)$ and $f_k(x)$ in (2.2).

We assume that the index of $b(x)$ on X-axis is equal to zero.

As it has been shown in section 2 the general solution of equation (2.4) is defined by (2.6). The formula (2.6) can be represented as follows

$$u(x,y) = e^{2i(\beta x - \alpha y)} \psi(z), \tag{2.119}$$

where

$$\psi(z) = \varphi(z) e^{-2i\beta z}. \tag{2.120}$$

If $u(x,y) \in M_0(R^2_+)$, then it follows from (2.119) that $\psi(z) \in M_0(R^2_+)$. Since $\psi(z)$ is analytic in R^2_+ we have $\psi(z) \in N_0(R^2_+)$ (cf. Remark 2.1).

Hence the general solution of equation (2.114) belonging to the class $M_0(R^2_+)$ is defined by

$$u(x,y) = u_0(x,y) + e^{2i(\beta x - \alpha y)} \psi(z), \tag{2.121}$$

where $\psi(z)$ is an arbitrary analytic in R^2_+ function belonging to the class $N_0(R^2_+)$ and $u_0(x,y)$ is the function from (2.115).

Substituting the general solution (2.121) into (2.118) we get

$$Re[b(x)\psi(x)] = f_0(x), \tag{2.122}$$

where

$$f_0(x) = f(x) - Re[e^{-2i\beta x} b(x) u_0(x,0)]. \tag{2.123}$$

As we have shown in section 2.4 the analytic function $\psi(z)$ is defined by the condition (2.122) uniquely (cf. section 2.4, problem A). Hence the problem (2.114), (2.118) in the class $M_0(R^2_+)$ is uniquely solvable.

Now we formulate a correct boundary value problem for equation (2.1) in general case.

Let $\lambda_1 = \alpha_1 + i\beta_1, \ldots, \lambda_n = \alpha_n + i\beta_n$ be the roots of equation

$$\lambda^n + a_1\lambda^{n-1} + \cdots + a_n = 0,$$

and let

$$(\lambda - \lambda_1)\ldots(\lambda - \lambda_k) \equiv \lambda^k + a_{k1}\lambda^{k-1} + \cdots + a_{kk}.$$

As the boundary conditions for (2.1) we take

$$Re[b_0(x)u(x,0)e^{-2i\beta_1 x}] = f_0(x), \quad x \in R^1, \tag{2.124}$$

$$Re\left[b_k(x)\left(\frac{\partial^k u(x,y)}{\partial \bar{z}^k} + a_{k1}\frac{\partial^{k-1}u(x,y)}{\partial \bar{z}^{k-1}} + \cdots + a_{kk}u(x,y)\right)e^{-2i\beta_{k+1}x}\right]$$
$$= f_k(x), \quad \text{for } y = 0,\ x \in R^1,\ k = 1, \ldots, n-1, \tag{2.125}$$

where $f_k(x)$ and $b_k(x)$ satisfy the same conditions as in (2.2). We assume that the index of $b_k(x)$ on X-axis is equal to zero.

One has the following

Theorem 2.8 *The problem (2.1), (2.124), (2.125) is uniquely solvable.*

We prove our assertion for $n = 2$. The general case can be treated similarly. Denote

$$w(x,y) = \frac{\partial u(x,y)}{\partial \bar{z}} - \lambda_1 u(x,y). \tag{2.126}$$

For $n = 2$ equation (2.1) takes the form

$$\frac{\partial w}{\partial \bar{z}} - \lambda_2 w = 0. \tag{2.127}$$

In view of (2.126) the condition (2.125) for $n = 2$ becomes

$$Re[e^{-2i\beta_2 x}b_1(x)w(x,0)] = f_1(x). \tag{2.128}$$

As we have shown above, the problem (2.127), (2.128) is uniquely solvable in the class $M_0(R^2_+)$ and the solution can be represented as

$$w = e^{2i(\beta_2 x - \alpha_2 y)}\psi_2(z), \tag{2.129}$$

where $\psi_2(z)$ is an analytic function from the class $N_0(R^2_+)$.

Substituting $w(x,y)$ from (2.129) into (2.126) we get the problem (2.124), (2.126) to determine a function $u(x,y)$ which is also uniquely solvable in the class $M_0(R^2_+)$.

Thus, for $n = 2$ the considered problem is uniquely solvable in the class $M_0(R^2_+)$.

2.9 Conclusions

Now let the functions $b_k(x)$ $(k = 0, 1, \ldots, n-1)$ satisfy the same conditions as in (2.2) and let the indices of these functions on X-axis be equal to zero. Let $f_k(x) \in N_\alpha(R^1)$ $(k = 0, 1, \ldots, n-1)$, where α is a positive number. Then using the above arguments we can prove that all the problems for equation (2.1) considered in this chapter are solvable in the class $M_\alpha(R^2_+)$ and the corresponding homogeneous problems in this class have exactly $n(m+1)$ linear independent solutions, where m is the entire part of α, if α is not an integer and $m = \alpha - 1$ otherwise.

The investigation of these problems in the classes $M_\alpha(R^2_+)$ is of particular interest when the limits of the functions $b_k(x)$ as $x \to +\infty$ and $x \to -\infty$ are different.

Chapter 3

Dirichlet Type Problem for the Product of First Order Differential Operators

3.1 Description of the Problem and Main Results

Let us consider the first order elliptic differential operators

$$L_j u = \mu_j \frac{\partial u}{\partial x} + i \frac{\partial u}{\partial t} + a_j u, \quad (j = 1, 2, \ldots, n), \tag{3.1}$$

where μ_j, a_j are constants, $Re\mu_j \neq 0$ and a_j, $j = 1, 2, \ldots, n$ are real.

Consider an elliptic equation of the form

$$L_1 L_2 \cdots L_n u = 0, \quad x \in R^1, \quad t > 0. \tag{3.2}$$

Equation (3.2) is a special case of equation (1.1), but it is more general than equation (2.1). In the case where $\mu_1 = \cdots = \mu_n = 1$, this equation coincides with equation (2.1). In chapter 1 we have investigated Dirichlet type problem for equation (1.1).

The aim of this chapter is to find an efficient method of resolution of this problem for equation (3.2). The solution of (3.2) is searched in the class $M_0(R^2_+)$ (the definitions of classes $N_\alpha(R^1)$ and $M_\alpha(R^1)$ are specified in section 1.1).

Observe that $Re\mu_j \neq 0$ $(j = 1, \ldots, n)$ is a condition of ellipticity of equation (3.2).

Let

$$Re\mu_j > 0, \quad j = 1, \ldots, r, \tag{3.3}$$

$$Re\mu_j < 0, \quad j = r+1, \ldots, n. \tag{3.4}$$

For the sake of definiteness we assume that $r \geq q$, where

$$q = n - r. \tag{3.5}$$

The boundary conditions are taken as

$$\frac{\partial^k u(x,0)}{\partial t^k} = f_k(x), \quad x \in R^1, \quad k = 0, \ldots, q-1, \tag{3.6}$$

$$Re\left[b_k(x)\frac{\partial^k u(x,0)}{\partial t^k}\right] = f_k(x), \quad x \in R^1, \quad k = q, \ldots, r-1, \tag{3.7}$$

where $f_k(x) \in N_0(R^1)$, $(k = 1, \ldots, r-1)$.

We assume that the functions $b_k(x)$ $(k = q, \ldots, r-1)$ satisfy the same conditions as in (2.2) and their indices on X-axis are equal to zero.
If $r = q$ the conditions (3.7) are absent and we get a Dirichlet problem for equation (3.2); if $q = 0$ the conditions (3.6) are absent and we get a Riemann-Hilbert problem.

The main result of this chapter is the following theorem.

Theorem 3.1 *The problem (3.2), (3,6), (3.7) in the class $M_0(R^2_+)$ is uniquely solvable.*

Below we will consider correct boundary value problems for equation (3.2), in the case where some of the constants μ_j are purely imaginary.

3.2 The General Solution of Non-Homogeneous Elliptic Equation

Consider a non-homogeneous first-order elliptic equation of the form

$$\mu\frac{\partial u}{\partial x} + i\frac{\partial u}{\partial t} + au = h(x,t), \quad t > 0, \quad x \in R^1, \tag{3.8}$$

where μ and a are constants, a is real and $Re\mu > 0$.

We assume that $h(z,t)$ is analytic on the first argument in $Imz > 0$ and satisfies the inequalities

$$\left|\frac{\partial^{k+j} h(z,t)}{\partial z^k \partial t^j}\right| \leq c_{kj}\frac{(1+\mid t \mid)^{\beta_{kj}}}{(1+\mid z \mid)^{\alpha_{kj}}}, \quad t \geq 0, \quad 0 \leq k+j < \infty, \tag{3.9}$$

where c_{kj}, β_{kj} and α_{kj} are some positive constants depending on $h(x,t)$.

Setting in (3.8)

$$u(x,t) = v(x,t)e^{iat} \tag{3.10}$$

we get

$$\mu\frac{\partial v}{\partial x} + i\frac{\partial v}{\partial t} = e^{-iat}h(x,t). \tag{3.11}$$

It is easy to check that a particular solution $v_0(x,t)$ of equation (3.11) can be determined by

$$v_0(x,t) = \int_0^t e^{-ia\tau}h(x+i\mu(t-\tau),\tau)d\tau. \tag{3.12}$$

As it is known (see [16]), the general solution of equation (3.11) at $h \equiv 0$ is determined by

$$v(x,t) = \varphi(x+i\mu t), \tag{3.13}$$

where $\varphi(z)$ is an arbitrary analytic in $Imz > 0$ function.

Hence the general solution of equation (3.8) is determined by

$$u(x,t) = \int_0^t e^{ai(t-\tau)}h(x+i\mu(t-\tau),\tau)d\tau + \varphi(x+i\mu t)e^{iat}, \tag{3.14}$$

where $\varphi(z)$ is an arbitrary analytic on complex variable $z = x+iy$ function in the half-plane $Imz > 0$.

Let $\mu = \mu_1 + i\mu_2$, $z = x+iy$, $y > 0$ and $\tau > 0$. By assumption $\mu_1 > 0$. Hence

$$| z + i\mu\tau |^2 = (x-\mu_2\tau)^2 + (y+\mu_1\tau)^2. \tag{3.15}$$

From here we get

$$| z+i\mu\tau |^2 \geq | z |^2 \quad \text{for} \quad \mu_2 = 0,$$
$$| z+i\mu\tau |^2 \geq \frac{| z |^2}{4} \quad \text{for} \quad | \mu_2 | \tau \leq \frac{| x |}{2}, \quad \mu_2 \neq 0,$$
$$| z+i\mu\tau |^2 \geq | y |^2 + \frac{{\mu_1}^2}{{\mu_2}^2}\frac{| x |^2}{4} \quad \text{for} \quad | \mu_2 | \tau \geq \frac{| x |}{2}, \quad \mu_2 \neq 0.$$

From these inequalities it follows that if $Re\mu > 0$, then

$$| z+i\mu\tau | \geq c_0 | z | \quad \text{for} \quad Imz \geq 0, \quad \tau \geq 0, \tag{3.16}$$

where c_0 is a constant depending only on μ. From (3.9) and (3.16) it follows that the function $v_0(z,t)$, defined by (3.12), also satisfies (3.9). Thus, $v_0 \in M_0(R^2_+)$.

Let the solution $u(x,t)$ of equation (3.8) belong to the class $M_0(R^2_+)$, then it follows from (3.14) that $\varphi(x+i\mu t)$ also belongs to the same class. Therefore $\varphi(z) \in N_0(R^2_+)$ (cf. Remark 2.1).

From here and from the inequality (3.16) it follows that $\varphi(z+i\mu t)$ satisfies the estimates of the form (3.9) with $\beta_{kj} = 0$.

Hence the general solution of equation (3.8) belonging to the class $M_0(R^2_+)$ is defined by (3.14), where $\varphi(z)$ is an arbitrary analytic in R^2_+ function, belonging to the class $N_0(R^2_+)$.

3.3 Cauchy Problem for Equation $L_1 \cdots L_r u = 0$

Let L_j be the operator defined by (3.1) with real constants a_j $(j = 1, \dots, r)$ and let $\mu_1, \dots, \mu_r$ satisfy the conditions (3.3). We consider the equation

$$L_1 \cdots L_r u = 0, \quad x \in R^1, \quad t > 0 \tag{3.17}$$

with Cauchy boundary conditions

$$\frac{\partial^k u(x,0)}{\partial t^k} = \psi_k(x), \quad x \in R^1, \quad k = 0, \dots, r-1, \tag{3.18}$$

where $\psi_k(z)$ are given analytic in R^2_+ functions belonging to the class $N_0(R^2_+)$.

One has the following

Theorem 3.2 *The Cauchy problem (3.17), (3.18) is uniquely solvable in the class $M_0(R^2_+)$.*

Proof. We set

$$w_0(x,t) = u(x,t), \tag{3.19}$$
$$w_k(x,t) = L_1 \cdots L_k u, \quad k = 1, \dots, r-1. \tag{3.20}$$

It is clear that

$$w_k(x,t) = \sum_{j+l \leq k} A_{kjl} \frac{\partial^{l+j} u(x,t)}{\partial x^l \partial t^j}, \tag{3.21}$$

where A_{kjl} are some constants and $A_{k0k} = (i)^k$.

From (3.18) we have

$$w_k(x,0) = \sum_{j+l \leq k} A_{kjl} \psi_j^{(l)}(x), \quad k = 0, \dots, r-1. \tag{3.22}$$

The inverse is also true: conditions (3.18) follow from the conditions (3.22). Hence (3.18) can be replaced by (3.22).

In view of (3.20), equation (3.17) for $k = r-1$ can be written as follows

$$\mu_r \frac{\partial w_{r-1}}{\partial x} + i\frac{\partial w_{r-1}}{\partial t} + a_r w_{r-1} = 0. \qquad (3.23)$$

Since $u \in M_0(R^2_+)$, the function w_{r-1} also belongs to the same class. Let us consider equation (3.23) with respect to w_{r-1} in the class $M_0(R^2_+)$. According to (3.14) the general solution of this equation in the class $M_0(R^2_+)$ is defined by

$$w_{r-1}(x,t) = \varphi_{r-1}(x + i\mu_r t)\exp(ia_r t) \qquad (3.24)$$

where $\varphi_{r-1}(z)$ is an arbitrary analytic in R^2_+ function from the class $M_0(R^2_+)$. From (3.24) and (3.22) at $k = r-1$ we get

$$\varphi_{r-1}(x) = \omega_{r-1}(x), \quad x \in R^1, \qquad (3.25)$$

where

$$\omega_{r-1}(z) = \sum_{j+l \le r-1} A_{r-1,jl}\psi_j^{(l)}(z), \quad Imz > 0. \qquad (3.26)$$

Since $\varphi_{r-1}(z)$ and $\omega_{r-1}(z)$ are analytic in R^2_+ functions, then by (3.25) we have

$$\varphi_{r-1}(z) = \omega_{r-1}(z), \quad Imz > 0. \qquad (3.27)$$

Substituting $w_{r-1}(x,t)$ from (3.24) into (3.20) at $k = r-1$ we get

$$L_1 \cdots L_{r-1} u = h_{r-2}(x,t), \qquad (3.28)$$

where

$$h_{r-2}(z,t) = \varphi_{r-1}(z + i\mu_r t)\exp(ia_r t) \qquad (3.29)$$

and $\varphi_{r-1}(z)$ is defined by (3.27).

In view of (3.20), equation (3.28) for $k = r-2$ can be written as follows

$$\mu_{r-1} \frac{\partial w_{r-2}}{\partial x} + i\frac{\partial w_{r-2}}{\partial t} + a_{r-1} w_{r-2} = h_{r-2}(x,t). \qquad (3.30)$$

According to (3.14) the general solution of equation (3.30) in the class $M_0(R^2_+)$ is defined by

$$\begin{aligned} w_{r-2}(x,t) &= \int_0^t e^{ia_{r-1}(t-\tau)} h_{r-2}(x + i\mu_{r-1}(t-\tau),\tau)\,d\tau \\ &+ \varphi_{r-2}(x + i\mu_{r-1}t)\exp(ia_{r-1}t), \end{aligned} \qquad (3.31)$$

where $\varphi_{r-2}(z)$ is an arbitrary analytic function from the class $M_0(R^2_+)$.

Setting $t = 0$ in (3.31) we get

$$w_{r-2}(x,0) = \varphi_{r-2}(x).$$

From here and (3.22) at $j = r-2$ we obtain

$$\varphi_{r-2}(z) = \sum_{j+k\leq r-2} A_{r-2,jl}\psi_j^{(l)}(z). \tag{3.32}$$

Thus, $L_1 \cdots L_{r-2}u = w_{r-2}(x,t)$, where $w_{r-2}(x,t)$ is defined by (3.31). Continuing this process and using the remaining conditions from (3.22) we uniquely determine a solution of the Cauchy problem (3.17), (3.18). From here the assertion of Theorem 3.2 follows immediately.

Theorem 3.2 is proved.

3.4 Riemann–Hilbert Problem for Equation (3.2)

Let $L_1, \ldots, L_n$ be the operators defined by (3.1), where constants a_j ($j = 1, \ldots, n$) are real and μ_j ($j = 1, \ldots, n$) satisfy conditions (3.3) and (3.4).

First we consider the equation

$$L_1 L_2 \cdots L_r u = 0, \quad t > 0, \quad x \in R^1. \tag{3.33}$$

One has the following

Lemma 3.1 *If $u(x,t)$ is a solution of equation (3.33) belonging to the class $M_0(R^2_+)$, then*

$$\frac{\partial^k u(x,0)}{\partial t^k} = \psi_k(x), \quad x \in R^1, \quad k = 0, 1, \ldots, r-1, \tag{3.34}$$

where $\psi_k(z)$ ($k = 0, 1, \ldots, r-1$) are some analytic in R^2_+ functions belonging to the class $N_0(R^2_+)$.

Proof. Let $u(x,t)$ be a solution of equation (3.33) and $u(x,t) \in M_0(R^2_+)$. Let $w_k(x,t)$ ($k = 0, 1, \ldots, r-1$) be the functions defined by (3.19) and (3.20). As we have shown in section 3.3 the function $w_{r-1}(x,t)$ can be represented as in (3.24), where $\varphi_{r-1}(z)$ is analytic in R^2_+ function, belonging to the class $N_0(R^2_+)$.

Hence we can write

$$w_{r-1}(x,0) = \varphi_{r-2}(x), \quad x \in R^1. \tag{3.35}$$

In a similar way, from (3.31) we have

$$w_{r-2}(x,0) = \varphi_{r-2}(x), \quad x \in R^1,$$

where $\varphi_{r-2}(z)$ is analytic in R^2_+ function belonging to the class $M_0(R^2_+)$.
Continuing this process we get

$$w_k(x,0) = \varphi_k(x), \quad x \in R^1, \quad k = 0,1,\ldots,r-1, \tag{3.36}$$

where $\varphi_0(z),\ldots,\varphi_{r-1}(z)$ are analytic in $Imz > 0$ functions belonging to the class $N_0(R^2_+)$.
From (3.21) and (3.36) we have

$$\sum_{j+l\le k} A_{kjl}\frac{\partial^{l+j}u(x,0)}{\partial x^l \partial t^j} = \varphi_k(x), \quad x \in R^1, \quad k = 0,1,\ldots,r-1. \tag{3.37}$$

From (3.37) at $k = 0$ we get

$$u(x,0) = \varphi_0(x). \tag{3.38}$$

Continuing the same process we find that the functions

$$\frac{\partial^k u(x,0)}{\partial t^k}, \quad k = 0,1,\ldots,r-1,$$

can be represented as in (3.34).
Lemma 3.1 is proved.
The boundary conditions for equation (3.33) will be taken as

$$Re\left[b_k(x)\frac{\partial^k u(x,0)}{\partial t^k}\right] = f_k(x), \quad x \in R^1, \quad k = 0,1,\ldots,r-1, \tag{3.39}$$

where $b_k(x)$ and $f_k(x)$ $(k = 0,1,\ldots,r-1)$ satisfy the same conditions as in (2.2). Here we also assume that the indices of $b_k(x)$ on X-axis are equal to zero.
One has the following

Theorem 3.3 *The problem (3.33), (3,39) is uniquely solvable in the class* $M_0(R^2_+)$.

Proof. In chapter 2 we have reduced the problem (2.1), (2.2) to the Cauchy problem for equation (2.1). In a similar way, using Lemma 3.1, the problem (3.33), (3.39) can be reduced to the Cauchy problem (3.17), (3.18) which, according to Theorem 3.2, is uniquely solvable.

3.5 The Existence of Solution of the Problem (3.2), (3.6), (3.7)

In this section we prove the existence of a solution for the problem (3.2), (3.6), (3.7) and construct this solution. In the next section we will prove the uniqueness of a solution for this problem.

Let the constants a_j $(j = 1, \dots, n)$ be real and $\mu_1, \dots, \mu_n$ satisfy the conditions (3.3), (3.4). We consider the operators

$$L_j^* v \equiv -\overline{\mu_j}\frac{\partial v}{\partial x} + i\frac{\partial v}{\partial t} - a_j v, \quad (j = 1, 2, \dots, n). \tag{3.40}$$

Let $v(x,t)$ be a solution of the equation $L_j^* v = 0$. It is clear that $u(x,t) = \overline{v(x,t)}$ is a solution of the equation $L_j u = 0$ where $\overline{v(x,t)}$ is the complex conjugate of $v(x,t)$.

Let us consider the equations

$$L_1 \cdots L_r v = 0, \quad x \in R^1, \quad t > 0, \tag{3.41}$$

$$L_{r+1}^* \cdots L_n^* w = 0, \quad x \in R^1, \quad t > 0. \tag{3.42}$$

It is easy to see that, if $v(x,t)$ and $w(x,t)$ are the solutions of equations (3.41) and (3.42), respectively and $v(x,t) \in M_0(R_+^2)$, $w(x,t) \in M_0(R_+^2)$ then $L_{r+1} \cdots L_n \overline{w} = 0$ and the function

$$u(x,t) = v(x,t) + \overline{w(x,t)} \tag{3.43}$$

is a solution of equation (3.2).

A solution of the problem (3.2), (3.6), (3.7) is searched in the form (3.43), where $v(x,t)$ and $w(x,t)$ are the solutions of equations (3.41) and (3.42), respectively, and belong to the class $M_0(R_+^2)$.

Substituting $u(x,t)$ from (3.43) into the boundary conditions (3.6) and (3.7) we get

$$\frac{\partial^k v(x,0)}{\partial t^k} + \frac{\partial^k \overline{w(x,0)}}{\partial t^k} = f_k(x), \quad k = 0, \dots, q-1, \tag{3.44}$$

$$Re\left[b_k(x)\left(\frac{\partial^k v(x,0)}{\partial t^k} + \frac{\partial^k \overline{w(x,0)}}{\partial t^k}\right)\right] = f_k(x), \quad k = q, \dots, r-1. \tag{3.45}$$

According to Lemma 3.1 the derivatives $\frac{\partial^k v(x,0)}{\partial t^k}$ and $\frac{\partial^k w(x,0)}{\partial t^k}$ can be represented as

$$\frac{\partial^k v(x,0)}{\partial t^k} = \varphi_k(x), \quad x \in R^1, \quad k = 0, 1, \dots, \tag{3.46}$$

$$\frac{\partial^k w(x,0)}{\partial t^k} = \psi_k(x), \quad x \in R^1, \quad k = 0, 1, \ldots, \tag{3.47}$$

where $\varphi_k(x)$ and $\psi_k(x)$ $(k = 0, 1, \ldots)$ are analytic in R^2_+ functions belonging to the class $N_0(R^2_+)$.

Substituting the values of $\frac{\partial^k v(x,0)}{\partial t^k}$ and $\frac{\partial^k w(x,0)}{\partial t^k}$ $(k = 0, 1, \ldots, r-1)$ from (3.46) and (3.47) into (3.44) we get

$$\varphi_k(x) + \overline{\psi_k(x)} = f_k(x), \quad x \in R^1, \quad k = 0, \ldots, q-1. \tag{3.48}$$

As we have shown in section 2.7 a solution of the conjugate problem (2.77) is defined by (2.83), (2.84). So a solution of the conjugate equation (3.48) is defined by

$$\varphi_k(z) = \frac{1}{2\pi i}\int_{-\infty}^{+\infty} \frac{f_k(t)dt}{t-z} \quad \text{and} \quad \psi_k(z) = \frac{1}{2\pi i}\int_{-\infty}^{+\infty} \frac{\overline{f_k(t)}dt}{t-z}, \tag{3.49}$$

where $Imz > 0$ and $k = 0, 1, \ldots, q-1$. Thus

$$\frac{\partial^k v(x,0)}{\partial t^k} = \varphi_k(x), \quad x \in R^1, \quad k = 0, \ldots, q-1, \tag{3.50}$$

$$\frac{\partial^k w(x,0)}{\partial t^k} = \psi_k(x), \quad x \in R^1, \quad k = 0, \ldots, q-1, \tag{3.51}$$

where $\varphi_k(x)$ and $\psi_k(x)$ are defined by (3.49).

Therefore, we get the Cauchy problem (3.42), (3.51) to determine $w(x,t)$, which in the class $M_0(R^2_+)$ is uniquely solvable (see Theorem 3.2). Taking into account (3.46), the condition (3.45) can be written as follows

$$Re[b_k(x)\varphi_k(x)] = g_k(x), \quad x \in R^1, \quad k = q, \ldots, r-1, \tag{3.52}$$

where

$$g_k(x) = f_k(x) - Re\left[b_k(x)\frac{\partial^k \overline{w(x,0)}}{\partial t^k}\right], \quad k = q, \ldots, r-1. \tag{3.53}$$

Since $w(x,t)$ is a known function from the class $M_0(R^2_+)$, then

$$g_k(x) \in N_0(R^1), \quad k = q, \ldots, r-1.$$

From the boundary condition (3.52) we find uniquely the analytic function $\varphi_k(z)$ $(k = q, \ldots, r-1)$ from the class $N_0(R^2_+)$. Hence

$$\frac{\partial^k v(x,0)}{\partial t^k} = \varphi_k(x), \quad x \in R^1, \quad k = 0, 1, \ldots, r-1, \tag{3.54}$$

where $\varphi_k(z)$ are known functions, belonging to the class $N_0(R^2_+)$.

Hence we again get the Cauchy problem (3.41), (3.54) to determine $u(x,t)$, which is uniquely solvable. Substituting $v(x,t)$ and $w(x,t)$ into (3.43) we get the searched solution of the problem (3.2), (3.6), (3,7).

Remark 3.1. We have proved, in fact, that the problem (3.2), (3.6), (3.7) is uniquely solvable in the class of functions admitting the representation (3.43). In the next section we will show that any solution of equation (3.2) belonging to the class $M_0(R^2_+)$ can be represented as in (3.43). This argument proves the uniqueness of a solution of the problem (3.2), (3.6), (3.7).

3.6 The Uniqueness of Solution of the Problem (3.2), (3.6), (3.7)

Let the constants a_j and μ_j $(j = 1,\ldots,n)$ be as in section 3.1, x_0 be a fixed point in R^1. Let L_k and L_k^* $(k = 1,\ldots,n)$ be the operators defined by (3.1) and (3.40), respectively. Consider the operators A_k and A_k^* $(k = 1,\ldots,n)$ defined by

$$A_kY(t) \equiv iY'(t) + (a_k - i\mu_k x_0)Y(t), \tag{3.55}$$

$$A_k^*Z(t) \equiv (a_k + i\bar{\mu}_k x_0)Z(t) - iZ'(t). \tag{3.56}$$

To prove the uniqueness of a solution of the problem (3.2), (3.6), (3.7) we consider the following two problems

Problem B_1.

$$A_1\cdots A_nY(t) = 0, \quad t > 0, \tag{3.57}$$

$$A_1^*\cdots A_n^*Z(t) = 0, \quad t > 0, \tag{3.58}$$

$$Y^{(k)}(0) = 0, \quad k = 0,\ldots,q-1, \tag{3.59}$$

$$Z^{(k)}(0) = 0, \quad k = 0,\ldots,q-1, \tag{3.60}$$

$$Y^{(k)}(0) + Z^{(k)}(0) = 0, \quad k = q,\ldots,r-1. \tag{3.61}$$

Problem B_2.

$$L_1L_2\cdots L_nu = 0, \quad x \in R^1, \quad t > 0,$$

$$L_1^*L_2^*\cdots L_n^*v = 0, \quad x \in R^1, \quad t > 0,$$

$$\frac{\partial^k u(x,0)}{\partial t^k} = 0, \quad x \in R^1, \quad k = 0,1,\ldots,q-1,$$

$$\frac{\partial^k v(x,0)}{\partial t^k} = 0, \quad x \in R^1, \quad k = 0,1,\ldots,q-1,$$

$$\frac{\partial^k u(x,0)}{\partial t^k} + \frac{\partial^k v(x,0)}{\partial t^k} = 0, \quad x \in R^1, \quad k = q,\ldots,r-1.$$

Recall that $q = n - r$ and $q \leq r$.

A solution of the problem B_1 is searched in the class of functions growing with their derivatives of an arbitrary order not more rapidly than a polynomial at infinity. A solution of the problem B_2 is searched in the class $M_0(R^2_+)$.

One has the following lemmas

Lemma 3.2 *For $x_0 \neq 0$ the problem B_1 has only zero solution.*

Lemma 3.3 *The problem B_2 has only zero solution.*

Proof of Lemma 3.2. Let $x_0 > 0$. We set

$$Y_1(t) = A_2 \cdots A_n Y. \tag{3.62}$$

Equation (3.57) takes the form

$$A_1 Y_1(t) = 0, \quad t > 0. \tag{3.63}$$

According to (3.55), equation (3.63) can be written as follows

$$Y_1'(t) = (\mu_1 x_0 + i a_1) Y_1(t). \tag{3.64}$$

The general solution of this equation is defined by

$$Y_1(t) = c_1 \exp(t \mu_1 x_0 + i a_1 t).$$

Since $x_0 > 0$ and $Re \mu_1 > 0$ then $Y_1(t)$ grows not more rapidly (as $t \to +\infty$) than a polynomial if and only if $c_1 = 0$, i.e. $Y_1(t) \equiv 0$. Substituting $Y_1(t) = 0$ into (3.62), we get

$$A_2 \cdots A_n Y(t) = 0, \quad t > 0. \tag{3.65}$$

Arguing in the same manner we get

$$A_{r+1} \cdots A_n Y(t) = 0, \quad t > 0. \tag{3.66}$$

In the same way from equation (3.58) we get

$$A_1^* \cdots A_r^* Z(t) = 0, \quad t > 0. \tag{3.67}$$

Equations (3.66) and (3.67) are ordinary differential equations of order q and r, respectively. From (3.66) and (3.59) it follows that $Y(t) \equiv 0$. Hence from (3.67), (3.60) and (3.61) it follows that $Z(t) \equiv 0$. The case $x_0 < 0$ can be considered in the same way.

Lemma 3.2 is proved.

Proof of lemma 3.3. According to the definition (cf. [8], p. 133), the problem B_2 satisfies the Lopatinsky condition at the point x_0 if the problem B_1 has only trivial solution. From Lemma 3.2 it follows that the problem B_2 satisfies the Lopatinsky condition at each point $x_0 \in R^1,\ x_0 \neq 0$. Hence from the result of the paper [8] (see p. 138, theorem 3) it follows that the problem B_2 does not have a non-zero solution.

Lemma 3.3 is proved.

Now we prove that any solution of equation (3.2) belonging to the class $M_0(R^2_+)$ can be represented as in (3.43). According to Remark 3.1 it is sufficient to prove the uniqueness of a solution of the problem (3.2), (3.6), (3.7).

Let $u_0(x,t)$ be a solution of equation (3.2) and $u_0(x,t) \in M_0(R^2_+)$.
We set

$$f_k(x) = \frac{\partial^k u_0(x,0)}{\partial t^k}, \quad k = 0,\dots,q-1,$$
$$f_k(x) = 2Re\frac{\partial^k u_0(x,0)}{\partial t^k}, \quad k = q,\dots,r-1.$$

Consider the problem (3.2), (3.6), (3.7) with the boundary conditions

$$\frac{\partial^k u(x,0)}{\partial t^k} = f_k(x), \quad x \in R^1, \quad k = 0,\dots,q-1, \tag{3.68}$$
$$Re\frac{\partial^k u(x,0)}{\partial t^k} = \frac{1}{2} f_k(x), \quad x \in R^1, \quad k = q,\dots,r-1. \tag{3.69}$$

We are going to prove that the homogeneous problem (3.2), (3.68), (3.69) (at $f_k = 0,\ k = 0,\dots,r-1$) has only a trivial solution.

Indeed, if $u(x,t)$ is a solution of the homogeneous problem (3.2), (3.68), (3.69) then $u(x,t)$ and $v(x,t) \equiv \overline{u(x,t)}$ are solutions of the problem B_2. Hence, according to Lemma 3.3, $u(x,t) = 0$.

Therefore, the non-homogeneous problem (3.2), (3.68), (3.69) is uniquely solvable. On the other hand, the function $u_0(x,t)$ is a solution of the problem (3.2), (3.68), (3.69).

As we have shown in the last section this problem has a solution $u_1(x,t)$ of the form (3.43). Since a solution of the problem (3.2), (3.68), (3.69) is unique we have $u_0(x,t) = u_1(x,t)$.

Thus, any solution $u_0(x,t)$ of equation (3.2) belonging to the class $M_0(R^2_+)$ can be represented as in (3.43). So the uniqueness of a solution of the problem (3.2), (3.6), (3.7) is proved.

Now, instead of the boundary conditions (3.7), we consider the boundary conditions

$$\sum_{j=1}^{k} Re\left[b_{kj}(x)\frac{\partial^j u(x,0)}{\partial t^j}\right] = f_k(x), \quad x \in R^1, \quad k = q,\dots,r-1, \tag{3.70}$$

where the functions $f_k(x)$ and $b_{kj}(x)$ belong to the class $N_0(R^1)$ and $b_{kk}(x)$ satisfy the same conditions as $b_k(x)$ in (3.7).

Using the arguments of the proof of Theorem 3.1, we can prove that the problem (3.2), (3.6) and (3.70) is also uniquely solvable in the class $M_0(R^2_+)$.

3.7 Dirichlet Problem for Second Order Elliptic Equations

Let L_1 and L_2 be first order elliptic operators defined by

$$L_1 u \equiv \mu_1 \frac{\partial u}{\partial x} + i \frac{\partial u}{\partial t} + a_1 u, \tag{3.71}$$

$$L_2 u \equiv \mu_2 \frac{\partial u}{\partial x} + i \frac{\partial u}{\partial t} + a_2 u, \tag{3.72}$$

where μ_1, μ_2, a_1, a_2 are complex constants satisfying

$$Re\mu_1 > 0, \quad Re\mu_2 < 0, \tag{3.73}$$

$$\frac{Ima_1}{Re\mu_1} = \frac{Ima_2}{Re\mu_2}. \tag{3.74}$$

Consider the equation

$$L_1 L_2 u = 0, \quad x \in R^1, \quad t > 0 \tag{3.75}$$

with the boundary condition

$$u(x, 0) = f(x), \quad x \in R^1, \tag{3.76}$$

where $f(x)$ is a given function from the class $N_0(R^1)$.

A solution of the problem (3.75), (3.76) is searched in the class of functions $M_0(R^2_+)$.

Note that the conditions (3.73), (3.74) are necessary for the correctness of the problem (3.75), (3.76) (cf. [3], p. 246, theorem 3).

For $n = 2$, equation (3.75) differs from (3.2) by the fact that in this case the constants a_1 and a_2 can take complex values.

One has the following

Theorem 3.4 *The problem (3.75), (3.76) is uniquely solvable in the class* $M_0(R^2_+)$.

The existence of a solution of the problem (3.75), (3.76).
Let

$$a_1 = \alpha_1 + i\beta_1,\ a_2 = \alpha_2 + i\beta_2,\ \mu_1 = \gamma_1 + i\delta_1,\ \mu_2 = \gamma_2 + i\delta_2, \tag{3.77}$$

where α_j, β_j, γ_j and δ_j $(j = 1,2)$ are real constants and $\gamma_1 > 0$, $\gamma_2 < 0$.

First we find the general solution of equation

$$\mu_1 \frac{\partial u}{\partial x} + i\frac{\partial u}{\partial t} + a_1 u = 0, \tag{3.78}$$

belonging to the class $M_0(R^2_+)$.

To this end we search a particular solution of equation (3.78) of the form

$$u = e^{i(b_1 x + k_1 t)}, \tag{3.79}$$

where b_1 and k_1 are real constants.

Substituting $u(x,t)$ from (3.79) into (3.78) we get

$$ib_1\mu_1 - k_1 + a_1 = 0 \tag{3.80}$$

to determine the real constants b_1 and k_1.

Substituting a_1 and μ_1 from (3.77) into (3.80) we get

$$i(b_1\gamma_1 + \beta_1) - (b_1\delta_1 + k_1 - \alpha_1) = 0. \tag{3.81}$$

From here we have

$$b_1\gamma_1 + \beta_1 = 0, \quad b_1\delta_1 + k_1 - \alpha_1 = 0. \tag{3.82}$$

Hence

$$b_1 = -\frac{\beta_1}{\gamma_1}, \quad k_1 = \alpha_1 + \frac{\beta_1\delta_1}{\gamma_1}. \tag{3.83}$$

Changing the variable in equation (3.78)

$$u(x,t) = v(x,t)e^{i(b_1 x + k_1 t)} \tag{3.84}$$

and substituting $u(x,t)$ from (3.84) into (3.78) we get

$$\mu_1 \frac{\partial v}{\partial x} + i\frac{\partial v}{\partial t} = 0 \tag{3.85}$$

to determine $v(x,t)$.

According to (3.14), for $h \equiv 0$ the general solution of equation (3.85) is defined by

$$v(x,t) = \varphi_1(x + i\mu_1 t), \tag{3.86}$$

where $\varphi_1(z)$ is an arbitrary analytic in R^2_+ function.

From (3.84) and (3.86) we get

$$u(x,t) = \varphi_1(x + i\mu_1 t)e^{i(b_1 x + k_1 t)}. \tag{3.87}$$

It follows from (3.87) that if $u(x,t) \in M_0(R^2_+)$, then $\varphi_1(z)$ also belong to the same class. Hence by Remark 2.1 we have $\varphi_1(z) \in N_0(R^2_+)$.

Therefore, the general solution of equation (3.78) belonging to the class $M_0(R^2_+)$ is defined by (3.87), where $\varphi_1(z)$ is arbitrary analytic in $Imz > 0$ function on complex variable z, $\varphi_1(z) \in N_0(R^2_+)$ and b_1 and k_1 are defined by (3.83).

Consider now the equation

$$-\overline{\mu}_2 \frac{\partial w}{\partial x} + i\frac{\partial w}{\partial t} - \overline{a}_2 w = 0. \tag{3.88}$$

The general solution of equation (3.88) belonging to the class $M_0(R^2_+)$ is defined by

$$w(x,t) = \varphi_2(x - i\overline{\mu}_2 t)e^{-i(b_2 x + k_2 t)}, \tag{3.89}$$

where

$$b_2 = -\frac{\beta_2}{\gamma_2}, \quad k_2 = \alpha_2 + \frac{\beta_2 \delta_2}{\gamma_2}, \tag{3.90}$$

and $\varphi_2(z)$ is an arbitrary analytic in the half-plane $Imz > 0$ function on complex variable z belonging to the class $N_0(R^2_+)$.

It is clear that if $u(x,t)$ is a solution of equation

$$\mu_2 \frac{\partial u}{\partial x} + i\frac{\partial u}{\partial t} + a_2 u = 0, \tag{3.91}$$

then $w(x,t) = \overline{u(x,t)}$ is a solution of equation (3.88). Hence the general solution of equation (3.91) belonging to the class $N_0(R^2_+)$ can be represented as follows

$$u(x,t) = \overline{w(x,t)}, \tag{3.92}$$

where $w(x,t)$ is defined by (3.89).

Now we turn to the proof of the existence of a solution of the problem (3.75), (3.76).

In view of (3.87), (3.89) and (3.92) we search for a solution of this problem as

$$u(x,y) = \varphi_1(x + i\mu_1 t)e^{i(b_1 x + k_1 t)} + \overline{\varphi_2(x - i\overline{\mu_2} t)}e^{i(b_2 x + k_2 t)}, \tag{3.93}$$

where $\varphi_1(z)$ and $\varphi_2(z)$ are unknown analytic in R^2_+ functions and $\varphi_j(z) \in N_0(R^2_+)$, $j = 1, 2$.

According to (3.74), (3.83) and (3.90) we have

$$b_1 = b_2. \tag{3.94}$$

Since the functions (3.87) and (3.92) are solutions of the equations $L_1 u = 0$ and $L_2 u = 0$, respectively, and the operators L_1 and L_2 commute, then it is evident that the function $u(x,t)$ defined by (3.93) is a solution of $L_1 L_2 u = 0$.

Substituting $u(x,t)$ from (3.93) into the boundary condition (3.76) and taking into account the equality (3.94) we get

$$\varphi_1(x) + \overline{\varphi_2(x)} = g(x), \quad x \in R^1, \tag{3.95}$$

where

$$g(x) = f(x)e^{-ib_1 x}.$$

The analytic functions $\varphi_1(z)$ and $\varphi_2(z)$ are defined uniquely from the condition (3.95) by (cf. section 2.7):

$$\varphi_1(z) = \frac{1}{2\pi i}\int_{-\infty}^{+\infty} \frac{g(\tau)d\tau}{\tau - z}, \qquad \varphi_2(z) = \frac{1}{2\pi i}\int_{-\infty}^{+\infty} \frac{\overline{g(\tau)}}{t - z} d\tau. \tag{3.96}$$

Hence

$$\varphi_1(x + i\mu_1 t) = \frac{1}{2\pi i}\int_{-\infty}^{+\infty} \frac{g(\tau)d\tau}{\tau - x - i\mu_1 t}, \quad t > 0, \tag{3.97}$$

$$\varphi_2(x - i\overline{\mu_2} t) = \frac{1}{2\pi i}\int_{-\infty}^{+\infty} \frac{\overline{g(\tau)}d\tau}{\tau - x + i\bar{\mu}_2 t}, \quad t > 0. \tag{3.98}$$

Substituting $\varphi_1(x + i\mu_1 t)$ and $\varphi_2(x - i\bar{\mu}_2 t)$ from (3.97) and (3.98) into (3.93) we obtain a solution of the problem (3.75), (3.76).

The uniqueness of a solution of the problem (3.75), (3.76).

Let x_0 be a fixed point in R^1. To prove the uniqueness of a solution of the problem (3.75), (3.76) we consider the following problem

$$(i\frac{d}{dt} + (a_1 - i\mu_1 x_0)E)(i\frac{d}{dt} + (a_2 - i\mu_2 x_0)E)Y(t) = 0, \ t > 0, \tag{3.99}$$

$$Y(0) = 0, \tag{3.100}$$

where E is the identity operator.

A solution of the problem (3.99), (3.100) is searched in the class of function growing with their derivatives not more rapidly than a polynomial at infinity.

It follows from (3.74) that if

$$x_0 \neq [Ima_1][Re\mu_1]^{-1},$$

then the problem (3.99), (3.100) has only trivial solution. This means that the problem (3.75), (3.76) satisfies the Lopatinsky condition at any point $x_0 \in R^1$ except, may be the point $x_0 = [Ima_1][Re\mu_1]^{-1}$ (cf. [8], p.133).

Hence from the result of [8] (see p.138, theorem 3) it follows that the homogeneous problem (3.75), (3.76) (at $f \equiv 0$) in the class $M_0(R^2_+)$ has only zero solution. Thus, the problem (3.75), (3.76) in the class $M_0(R^2_+)$ is uniquely solvable.

Theorem 3.4 is proved.

3.8 Dirichlet Problem For 2m-th Order Elliptic Equations

Let $L_1, \ldots, L_{2m}$ be elliptic operators defined by

$$L_j u \equiv \mu_j \frac{\partial u}{\partial x} + i\frac{\partial u}{\partial t} + a_j u, \quad j = 1, \ldots, 2m, \tag{3.101}$$

where μ_j and a_j $(j = 1, \ldots, 2m)$ are complex constants, satisfying the conditions

$$Re\mu_j > 0, \quad j = 1, \ldots, m, \tag{3.102}$$

$$Re\mu_j < 0, \quad j = m+1, \ldots, 2m, \tag{3.103}$$

$$\frac{Ima_j}{Re\mu_j} = x_0, \quad j = 1, 2, \ldots, 2m, \tag{3.104}$$

and x_0 is a constant independent of j.

Let us consider the equation

$$L_1 \cdots L_{2m} u = 0, \quad x \in R^1, \quad t > 0 \tag{3.105}$$

with Dirichlet boundary conditions

$$\frac{\partial^k u(x,0)}{\partial t^k} = f_k(x), \quad k = 0, \ldots, m-1, \tag{3.106}$$

where $f_k(x)$ are given functions from the class $N_0(R^1)$.

A solution of the problem (3.105), (3.106) is searched in the class of function $M_0(R^2_+)$.

From the result of [3] (see p. 246, theorem 3) it follows that the conditions (3.102) and (3.103) are necessary for the correctness of the problem (3.105), (3.106).

In section 3.7 we have shown that for $n = 2$ the condition (3.104) is also necessary for the correctness of this problem.

One has the following

Theorem 3.5 *The problem (3.105), (3.106) in the class $M_0(R^2_+)$ is uniquely solvable.*

Proof. Let

$$a_j = \alpha_j + i\beta_j, \quad \mu_j = \gamma_j + i\delta_j \quad (j = 1, \ldots, 2m),$$

where α_j, β_j, γ_j and δ_j are real constants.

From (3.102) – (3.104) we have

$$\gamma_j > 0, \quad j = 1, \ldots, m, \tag{3.107}$$

$$\gamma_j < 0, \quad j = m+1, \ldots, 2m, \tag{3.108}$$

$$\frac{\beta_j}{\gamma_j} = x_0, \quad j = 1, 2, \ldots, 2m. \tag{3.109}$$

It is easy to verify the identity

$$L_j(v(x,t)e^{-x_0xi}) \equiv (\mu_j \frac{\partial v}{\partial x} + i\frac{\partial v}{\partial t} + b_j v)e^{-x_0xi}, \tag{3.110}$$

where

$$b_j = x_0\delta_j + \alpha_j. \tag{3.111}$$

Let Q_j be an operator defined by

$$Q_j v \equiv \mu_j \frac{\partial v}{\partial x} + i\frac{\partial v}{\partial t} + b_j v. \tag{3.112}$$

From (3.110) we have

$$L_1 \cdots L_{2m}(v(x,t)e^{-x_0xi}) = e^{-x_0xi}Q_1 \cdots Q_{2m}v. \tag{3.113}$$

Changing the variable

$$u(x,t) = e^{-x_0xi}v(x,t) \tag{3.114}$$

in equation (3.105) and in the boundary conditions (3.106) and taking into account (3.113), we get

$$Q_1 \cdots Q_{2m}v(x,t) = 0, \quad x \in R^1, \quad t > 0, \tag{3.115}$$

$$\frac{\partial^k v(x,0)}{\partial t^k} = e^{x_0xi}g_k(x), \quad k = 0, \ldots, m-1, \tag{3.116}$$

where $g_k(x)$ is a linear function of $f_0(x), \ldots, f_k(x)$. Since $u(x,t) \in M_0(R^2_+)$, then $v(x,t)$ is also searched in the same class.

The constants b_j are real, hence for $n = 2m, r = m, q = m$ the coefficients of equation (3.115) satisfy all the conditions of equation (3.2). Thus, according to Theorem 3.1, the problem (3.115), (3.116) is uniquely solvable.

Theorem 3.5 is proved.

The problem (3.75), (3.76) is a special case of the problem (3.105), (3.106) at $m = 2$. But in section 3.7 we have mentioned a simple method of resolution of the problem (3.75), (3.76).

3.9 Riemann–Hilbert Problem for Paired Elliptic Equation

Let L_j $(j = 1, \ldots, n)$ be the operators defined by (3.1), where the coefficients a_j $(j = 1, \ldots, n)$ are real constants and the constants μ_j $(j = 1, \ldots, n)$ satisfy (3.3), (3.4).

Let us consider the following two elliptic equations

$$L_1 \cdots L_r u = 0, \quad x \in R^1, \quad t > 0, \tag{3.117}$$

$$L_{r+1} \cdots L_n v = 0, \quad x \in R^1, \quad t > 0. \tag{3.118}$$

For the sake of definiteness, we assume that $r \geq q$ where $q = n - r$.

As the boundary conditions for equations (3.117), (3.118) we take

$$\frac{\partial^k u(x,0)}{\partial t^k} = b_k(x)\frac{\partial^k v(x,0)}{\partial t^k} + f_k(x), \; k = 0, \ldots, q-1, \; x \in R^1, \tag{3.119}$$

$$Re\left[b_k(x)\frac{\partial^k u(x,0)}{\partial t^k}\right] = f_k(x), \qquad k = q, \ldots, r-1, \quad x \in R^1. \tag{3.120}$$

We assume that the functions $b_k(x)$ and $f_k(x)$ $(k = 0, \ldots, r-1)$ satisfy the same conditions as in (3.6), (3.7). We also suppose that the indices of $b_k(x)$ on X-axis are equal to zero.

A solution of the problem (3.117) – (3.120) is searched in the class $M_0(R^2_+)$.

One has the following

Theorem 3.6 *The problem (3.117) – (3.120) in the class $M_0(R^2_+)$ is uniquely solvable.*

Proof. Let $\phi^+(z)$ and $\psi^-(z)$ be functions analytic in $Imz > 0$ and $Imz < 0$, respectively. Denote by $\phi^+(x)$ and $\psi^-(x)$ the limits of the functions $\phi^+(z)$ and $\psi^-(z)$ when $z \to x$, $Imz > 0$ and $z \to x$, $Imz < 0$, respectively.

First we consider the case where

$$b_k(x) = \frac{\phi_k^+(x)}{\phi_k^-(x)}, \quad x \in R^1, \quad k = 0, 1, \ldots, q-1, \tag{3.121}$$

$$b_k(x) = \phi_k^+(x), \quad x \in R^1, \quad k = q, \ldots, r-1, \tag{3.122}$$

where $\phi_k^+(z)$ and $\phi_j^-(z)$ ($k = 1, \ldots, r-1$, $j = 1, \ldots, q-1$) are analytic functions in the domains $Imz > 0$ and $Imz < 0$, respectively, and satisfy the inequalities

$$\phi^+(z) \neq 0, \qquad Imz \geq 0, \tag{3.123}$$

$$\phi^-(z) \neq 0, \qquad Imz \leq 0, \tag{3.124}$$

and

$$|\phi^+(z)|^{-1} \leq c, \quad \left|\frac{d^k \phi^+(z)}{dz^k}\right| \leq c_k, \quad Imz \geq 0, \ k = 0, 1, \ldots, \tag{3.125}$$

$$|\phi^-(z)|^{-1} \leq c, \quad \left|\frac{d^k \phi^-(z)}{dz^k}\right| \leq c_k, \quad Imz \leq 0, \ k = 0, 1, \ldots, \tag{3.126}$$

where c and c_k ($k = 0, 1, \ldots$) are some positive constants.

According to Lemma 3.1, if $u(x,t)$ is a solution of equation (3.117) and $u(x,t) \in M_0(R_+^2)$, then

$$\frac{\partial^k u(x,0)}{\partial t^k} = \omega_k^+(x), \quad x \in R^1, \quad k = 0, 1, \ldots, r-1. \tag{3.127}$$

where $\omega_k^+(z)$ is an analytic in R_+^2 function and $\omega_k^+(z) \in N_0(R_+^2)$.

Using the arguments of the proof of Lemma 3.1 we can prove that if $v(x,t)$ is a solution of equation (3.118) and $v(x,t) \in M_0(R_+^2)$, then

$$\frac{\partial^k v(x,0)}{\partial t^k} = \omega_k^-(x), \quad x \in R^1, \quad k = 0, 1, \ldots, q-1, \tag{3.128}$$

where $\omega_k^-(z)$ are some analytic in $Imz < 0$ functions and $\omega_k^-(z) \in N_0(R_-^2)$.

Substituting the values of

$$b_k(x) \ (k = 0, \ldots, r), \qquad \frac{\partial^k u(x,0)}{\partial t^k} \quad \text{and} \quad \frac{\partial^k v(x,0)}{\partial t^k}$$

from (3.121), (3.122), (3.127) and (3.128) into the boundary conditions (3.119) and (3.120), we get

$$\frac{\omega_k^+(x)}{\phi^+(x)} = \frac{\omega_k^-(x)}{\phi^-(x)} + \frac{f_k(x)}{\phi^+(x)}, \quad k = 0, \ldots, q-1, \quad x \in R^1, \tag{3.129}$$

$$Re[\phi^+(x)\omega_k^+(x)] = f_k(x), \quad k = q, \ldots, r-1, \quad x \in R^1. \tag{3.130}$$

Thus, we have obtained a well-known Riemann–Hilbert problem to determine the analytic functions $\omega_k^+(z)$ $(k = 0, \ldots, r-1)$ and $\omega_k^-(z)$ $(k = 0, \ldots, q-1)$ (see [15], pp. 136 and 187). A solution of this problem is defined by

$$\frac{\omega_k^+(z)}{\phi^+(z)} = \frac{1}{2\pi i}\int_{-\infty}^{+\infty} \frac{f_k(\zeta)}{\phi^+(\zeta)}.\frac{d\zeta}{\zeta - z}, \quad Imz > 0, \quad k = 0, 1, \ldots, q-1, \tag{3.131}$$

$$\frac{\omega_k^-(z)}{\phi_k^-(z)} = \frac{1}{2\pi i}\int_{-\infty}^{+\infty} \frac{f_k(\zeta)}{\phi^+(\zeta)}.\frac{d\zeta}{\zeta - z}, \quad Imz < 0, \quad k = 0, 1, \ldots, q-1, \tag{3.132}$$

$$\phi^+(z)\omega_k^+(z) = \frac{1}{\pi i}\int_{-\infty}^{+\infty} \frac{f_k(\zeta)d\zeta}{\zeta - z}, \quad Imz > 0, \quad k = q, \ldots, r-1. \tag{3.133}$$

So we have obtained the Cauchy problem (3.117), (3.127) and (3.118), (3.128) to determine $u(x,t)$ and $v(x,t)$ where the analytic functions $\omega_k^+(z)$ and $\omega_j^-(z)$ are defined by (3.131), (3.132) and (3.133). According to Theorem 3.2, this problem is uniquely solvable.

Now let us consider the general case. We are going to show that this problem in the general case can be reduced to the considered case (see [15], pp. 160–161).

Let

$$\phi_k^+(z) = b_k(\infty)\exp\left(\frac{1}{2\pi i}\int_{-\infty}^{+\infty} ln\frac{b_k(\zeta)}{b_k(\infty)}.\frac{d\zeta}{\zeta - z}\right), \quad Imz > 0, \tag{3.134}$$

$$\phi_k^-(z) = \exp\left(\frac{1}{2\pi i}\int_{-\infty}^{+\infty} ln\frac{b_k(\zeta)}{b_k(\infty)}.\frac{d\zeta}{\zeta - z}\right), \quad Imz < 0, \tag{3.135}$$

where $ln(b_k(\zeta)[b_k(\infty)]^{-1})$ is the branch of logarithm satisfying $ln 1 = 0$.

From Sokhotzky-Plemelj formula (see [15], p. 66) we have

$$\frac{\phi_k^+(x)}{\phi_k^-(x)} = b_k(x), \quad k = 0, 1, \ldots, q-1. \tag{3.136}$$

In section 2.4 the boundary condition (2.30) have been written in the form (2.38). In a similar way, the boundary conditions (3.120) can be written in the form

$$Re[\phi_k^+(x)\frac{\partial^k u(x,0)}{\partial t^k}] = g_k(x), \quad k = q, \ldots, r-1, \tag{3.137}$$

where $g_k(x) \in N_0(R^1_+)$ $(k = q, \ldots, r-1)$ and $\phi^+_k(z)$ $(k = q, \ldots, r-1)$ are analytic in $Imz > 0$, functions satisfying (3.123) and (3.125).

Thus, (3.117) – (3.120) have been reduced to the above considered case.

Theorem 3.6 is proved.

3.10 Dirichlet Problem for Generalized Analytic Functions

Consider the equation

$$\frac{\partial u}{\partial \bar{z}} + au + b\bar{u} = 0, \quad x \in R^1, \quad t > 0, \tag{3.138}$$

where a and b are complex constants and $b \neq 0$, $\bar{u}$ is the complex conjugate of u,

$$\frac{\partial}{\partial \bar{z}} \equiv \frac{1}{2}\left(\frac{\partial}{\partial x} + i\frac{\partial}{\partial t}\right).$$

Note that the boundary value problems for equation (3.138) at $b = 0$ are considered in chapter 2.

It has been shown in monograph [17] that any elliptic system of two first order equations with real coefficients can be reduced to equation (3.138). Many properties of analytic functions are valid for solutions of equation (3.138). For this reason the solution of equation (3.138) in monograph [17] has been called a generalized analytic function.

In the mentioned monograph correct boundary value problems in the bounded domains with arbitrary complex-valued functions a and b from the class L_p, $p > 2$ have been investigated. In particular, it has been shown that equation (3.138) with the boundary condition

$$Reu(x,y) = f(x,y), \quad (x,y) \in \Gamma, \quad Imu(x_0,y_0) = c_0, \tag{3.139}$$

in any simple connected bounded domain D is uniquely solvable. Here Γ is the boundary of the domain D, $f(x,y)$ is a real function defined on Γ, (x_0,y_0) is a fixed point of $D \cup \Gamma$ and c_0 is a given real number.

Let us consider equation (3.138) with the boundary condition

$$Reu(x,0) = f(x), \quad x \in R^1, \tag{3.140}$$

where $f(x)$ is a given real-valued function from the class $N_0(R^1)$.

We are going to show that the class of the function to be found, for which the problem (3.138), (3.140) is uniquely solvable essentially depends on the coefficients a and b of equation (3.138).

Definition 3.1. We say that a function $u(x,t)$ defined in the half-plane $x \in R^1$, $t \geq 0$ belongs to the class $M_0^\alpha(R^2_+)$ if $u(x,t)e^{-\alpha t} \in M_0(R^2_+)$.

Theorem 3.7 *If $Re a \neq Re b$, then for $\alpha = -2 Im a$ the problem (3.138), (3.140) is uniquely solvable in the class $M_0^\alpha(R_+^2)$.*

We prove Theorem 3.7 under the assumption that a is real. The general case can be reduced to this case by substituting $u(x,t) = e^{\alpha t} v(x,t)$, where $\alpha = -2 Im a$.

Passing to the complex conjugate function in equation (3.138) we get

$$\frac{\partial \bar{u}}{\partial z} + \bar{a}\bar{u} + \bar{b}u = 0, \tag{3.141}$$

where

$$\frac{\partial}{\partial z} = \frac{1}{2}(\frac{\partial}{\partial x} - i\frac{\partial}{\partial t}).$$

The boundary conditions (3.140) can be rewritten as

$$u(x,0) + \overline{u(x,0)} = 2f(x), \quad x \in R^1. \tag{3.142}$$

Denote

$$v(x,t) = u(x,t), \quad w(x,t) = \overline{u(x,t)}. \tag{3.143}$$

In these notations equations (3.138), (3.141) and the boundary condition (3.142) take the form

$$\frac{\partial v}{\partial \bar{z}} + av + bw = 0, \quad x \in R^1, \quad t > 0, \tag{3.144}$$

$$\frac{\partial w}{\partial z} + \bar{a}w + \bar{b}v = 0, \quad x \in R^1, \quad t > 0, \tag{3.145}$$

$$v(x,0) + w(x,0) = 2f(x), \quad x \in R^1. \tag{3.146}$$

Consider now the problem (3.144) – (3.146) with respect to the functions $v(x,t)$ and $w(x,t)$ in the class $M_0(R_+^2)$. If $v(x,t)$ and $w(x,t)$ are the solutions of the problem (3.144) – (3.146), then the function

$$u(x,t) = \frac{v(x,t) + \overline{w(x,t)}}{2} \tag{3.147}$$

is a solution of the problem (3.138), (3.140).

First let

$$|\, b \,|>|\, a \,|, \quad Re b \neq a, \tag{3.148}$$

and $f(x) \in C_0^\infty(R^1)$.

Under these hypotheses a solution of the problem (3.144) – (3.146) will be searched in the class $M_{-1}(R^2_+)$. Let $\varphi(\sigma,t)$ and $\psi(\sigma,t)$ be the Fourier transforms of $v(x,t)$ and $w(x,t)$ with respect to x, i.e.

$$\varphi(\sigma,t) = \frac{1}{\sqrt{2\pi}}\int\limits_{-\infty}^{+\infty} v(x,t)e^{ix\sigma}dx, \tag{3.149}$$

$$\psi(\sigma,t) = \frac{1}{\sqrt{2\pi}}\int\limits_{-\infty}^{+\infty} w(x,t)e^{ix\sigma}dx, \tag{3.150}$$

and let $v(x,t)$ and $w(x,t)$ belong to the class $M_{-1}(R^2_+)$.

For $t \geq 0$ and $\sigma \in R^1$ we have

$$\left| \frac{\partial^k \psi(\sigma,t)}{\partial t^k} \right| + \left| \frac{\partial^k \varphi(\sigma,t)}{\partial t^k} \right| \leq c_k(1+t)^{\beta_k}, \quad k = 0,1,\ldots \tag{3.151}$$

where c_k and β_k are some non-negative constants.

Calculating in (3.144), (3.145) and (3.146) the Fourier transforms with respect to x we get

$$i\frac{\partial\varphi(\sigma,t)}{\partial t} + (2a - i\sigma)\varphi(\sigma,t) + 2b\psi(\sigma,t) = 0, \tag{3.152}$$

$$-i\frac{\partial\psi(\sigma,t)}{\partial t} + (2a - i\sigma)\psi(\sigma,t) + 2\bar{b}\varphi(\sigma,t) = 0, \tag{3.153}$$

$$\varphi(\sigma,0) + \psi(\sigma,0) = 2F(\sigma), \tag{3.154}$$

where $\sigma \in R^1$, $t > 0$ and

$$F(\sigma) = \frac{1}{\sqrt{2\pi}}\int\limits_{-\infty}^{+\infty} f(x)e^{ix\sigma}dx. \tag{3.155}$$

From (3.152) we have

$$\psi(\sigma,t) = -\frac{i}{2b}\frac{\partial\varphi(\sigma,t)}{\partial t} + \frac{1}{2b}(i\sigma - 2a)\varphi(\sigma,t). \tag{3.156}$$

Substituting $\varphi(\sigma,t)$ from (3.156) into (3.153) we get

$$\frac{\partial^2\varphi(\sigma,t)}{\partial t^2} - (\sigma^2 + 4|\, b\,|^2 + 4ai\sigma - 4a^2)\varphi(\sigma,t) = 0. \tag{3.157}$$

Let

$$\lambda(\sigma) = \sqrt{\sigma^2 + 4|\, b\,|^2 + 4ai\sigma - 4a^2}. \tag{3.158}$$

Here by square root we mean the branch of continuous function with non-negative real part. If $| b |>| a |$, then $\lambda(\sigma)$ is infinitely differentiable on σ and

$$Re\lambda(\sigma) > 0 \quad for \quad \sigma \quad \in R^1. \tag{3.159}$$

We set

$$\omega(\sigma,t) = \frac{\partial\varphi(\sigma,t)}{\partial t} + \lambda(\sigma)\varphi(\sigma,t). \tag{3.160}$$

Then equation (3.157) takes the form

$$\frac{\partial\omega(\sigma,t)}{\partial t} - \lambda(\sigma)\omega(\sigma,t) = 0, \quad \sigma \in R^1, \quad t \geq 0. \tag{3.161}$$

The general solution of equation (3.161) is defined by

$$\omega(\sigma,t) = c_1(\sigma)e^{\lambda(\sigma)t}. \tag{3.162}$$

From (3.160) and (3.151) we have

$$| \omega(\sigma,t) |\leq c(1+ | \sigma |)(1+t)^{\beta}, \quad t > 0, \quad \sigma \in R^1, \tag{3.163}$$

where c and β are some constants.

Substituting $\omega(x,t)$ from (3.162) into (3.163) we get

$$| c_1(\sigma) |\leq c(1+ | \sigma |)(1+t)^{\beta}e^{-Re\lambda(\sigma)t}, \quad t > 0, \quad \sigma \in R^1. \tag{3.164}$$

Passing to the limit in (3.164) as $t \to +\infty$ and taking into account (3.159) we get $c_1(\sigma) = 0$ for $\sigma \in R^1$.

Hence

$$\omega(\sigma,t) = 0, \quad \sigma \in R^1, \quad t \geq 0. \tag{3.165}$$

Substituting $\omega(\sigma,t)$ from (3.165) into (3.160) we get

$$\frac{\partial\varphi(\sigma,t)}{\partial t} + \lambda(\sigma)\varphi(\sigma,t) = 0. \tag{3.166}$$

The general solution of equation (3.166) is defined by

$$\varphi(\sigma,t) = c_2(\sigma)e^{-\lambda(\sigma)t}, \tag{3.167}$$

where $c_2(\sigma)$ is an arbitrary function.

Substituting $\varphi(\sigma,t)$ from (3.167) into (3.156) we get

$$\psi(\sigma,t) = \frac{c_2(\sigma)}{2b}(i\sigma - 2a + i\lambda(\sigma))e^{-\lambda(\sigma)t}. \tag{3.168}$$

Substituting $\varphi(\sigma,t)$ and $\psi(\sigma,t)$ from (3.167) and (3.168) into the condition (3.154) we get

$$c_2(\sigma)(i\sigma - 2a + i\lambda(\sigma) + 2b) = 4bF(\sigma). \tag{3.169}$$

Let us prove that for $Reb \neq a$ and $\sigma \in R^1$

$$i\sigma - 2a + i\lambda(\sigma) + 2b \neq 0. \tag{3.170}$$

To this end we consider the equation

$$i\sigma - 2a + i\lambda(\sigma) + 2b = 0 \tag{3.171}$$

with respect to the real variable σ.

Equation (3.171) can be rewritten as

$$\lambda(\sigma) = -\sigma - 2i(a-b). \tag{3.172}$$

From (3.172) we have

$$\lambda^2(\sigma) = (-\sigma - 2i(a-b))^2. \tag{3.173}$$

Substituting $\lambda(\sigma)$ from (3.158) into (3.173) we get

$$| b |^2 = -i\sigma b - b^2 + 2ab. \tag{3.174}$$

From (3.172) we obtain

$$\sigma = \bar{b}i + ib - 2ai. \tag{3.175}$$

Hence

$$\sigma = (2Reb - 2a)i. \tag{3.176}$$

Since $a \neq Reb$, equation (3.176) has no roots with respect to the real variable σ. Hence equation (3.171) also has no real roots and we obtain the relation (3.170).

From (3.167), (3.168) and (3.169) we have

$$\varphi(\sigma,t) = \omega_1(\sigma)e^{-\lambda(\sigma)t}F(\sigma), \tag{3.177}$$

$$\psi(\sigma,t) = \omega_2(\sigma)e^{-\lambda(\sigma)t}F(\sigma), \tag{3.178}$$

where

$$\omega_1(\sigma) = \frac{4b}{i\sigma - 2a + i\lambda(\sigma) + 2b}, \tag{3.179}$$

$$\omega_2(\sigma) = \frac{2(i\sigma - 2a + i\lambda(\sigma))}{i\sigma - 2a + i\lambda(\sigma) + 2b}. \tag{3.180}$$

Since $\varphi(\sigma,t)$ and $\psi(\sigma,t)$ are the Fourier transforms of $v(x,t)$ and $w(x,t)$ with respect to x, then from (3.177) and (3.178) we have

$$v(x,t) = \frac{1}{\sqrt{2\pi}} \int_{-\infty}^{+\infty} \omega_1(\sigma) e^{-\lambda(\sigma)t} F(\sigma) e^{-i\sigma x} d\sigma, \tag{3.181}$$

$$w(x,t) = \frac{1}{\sqrt{2\pi}} \int_{-\infty}^{+\infty} \omega_2(\sigma) e^{-\lambda(\sigma)t} F(\sigma) e^{-i\sigma x} d\sigma. \tag{3.182}$$

Since $\mid b \mid > \mid a \mid$ and $Reb \neq a$, then from (3.158) we obtain

$$\lim_{\sigma \to -\infty} (i\sigma - 2a + i\lambda(\sigma) + 2b) = 2b, \tag{3.183}$$

$$\lim_{\sigma \to +\infty} \frac{1}{2i\sigma} [i\sigma - 2a + i\lambda(\sigma) + 2b] = 1. \tag{3.184}$$

Since $F(\sigma)$ is the Fourier transform of an infinitely differentiable and finite function $f(x)$, then (see [3], p. 138)

$$\lim_{|\sigma| \to \infty} \sigma^j \frac{d^k F(\sigma)}{d\sigma^k} = 0, \quad j,\, k = 0, 1, 2, \ldots. \tag{3.185}$$

From the relations (3.170), (3.183), (3.184) and (3.185) it follows that the functions (3.181) and (3.182) belong to the class $M_{-1}(R^2_+)$.

Thus, under our hypotheses on a, b and $f(x)$ the function

$$u(x,t) = \frac{1}{2}(v(x,t) + \overline{w(x,t)}) \tag{3.186}$$

is a solution of the problem (3.138), (3.140) in the class $M^0_{-1}(R^2_+)$, where $v(x,t)$ and $w(x,t)$ are defined by (3.181) and (3.182).

Substituting $F(\sigma)$ from (3.155) into (3.181) and (3.182) and changing the order of integrals we get

$$v(x,t) = \int_{-\infty}^{+\infty} \phi_1(x-\zeta,t) f(\zeta) d\zeta, \tag{3.187}$$

$$w(x,t) = \int_{-\infty}^{+\infty} \phi_2(x-\zeta,t) f(\zeta) d\zeta, \tag{3.188}$$

where

$$\phi_j(x,t) = \frac{1}{2\pi}\int_{-\infty}^{+\infty} \omega_j(\sigma)e^{-\lambda(\sigma)t}e^{-i\sigma x}d\sigma. \tag{3.189}$$

Now assume that a be real, $Reb \neq a$ and $f(x) \in N_0(R^1)$. We prove that the functions $v(x,t)$ and $w(x,t)$ defined by (3.187), (3.188) belong to the class $M_0(R^2_+)$ and satisfy the problem (3.144) – (3.146). Hence the function (3.186) is a solution of the problem (3.138), (3.140).

Arguing as in the proof of Lemma 3.3 we can prove that the homogeneous problem (3.144) – (3.146) in the class $M_0(R^2_+)$ has only trivial solution.

Thus, for $Rea \neq Reb$ the problem (3.38), (3.140) in the class $M_0^\alpha(R^2_+)$ is uniquely solvable.

Theorem 3.7 is proved.

Now let a be real and $Reb = a$. In this case the problem (3.138), (3.140) needs a special investigation, since the denominators of (3.179) and (3.180) at the point $\sigma = 0$ are equal to zero.

On the basis of (3.181), (3.182) it seems necessary to impose on $f(x)$ the additional restriction

$$\int_{-\infty}^{+\infty} f(x)dx = 0. \tag{3.190}$$

3.11 Correct Boundary Value Problems for Products of First Order Differential Operators

Let us consider the first order operators

$$L_j u \equiv \mu_j\frac{\partial u}{\partial x} + i\frac{\partial u}{\partial t} + a_j u, \quad (j = 1,\ldots,n). \tag{3.191}$$

We assume that

$$Re\mu_j > 0, \quad j = 1,\ldots,r_1, \tag{3.192}$$
$$Re\mu_j < 0, \quad j = r_1+1,\ldots,r_2, \tag{3.193}$$
$$Re\mu_j = 0, \quad j = r_2+1,\ldots,n, \tag{3.194}$$
$$Ima_j = 0, \quad j = 1,\ldots,r_2, \tag{3.195}$$
$$Ima_j \geq 0, \quad j = r_2+1,\ldots,n. \tag{3.196}$$

Denote by q, ρ, q_0, r and p the following integers

$$q = \min(r_1, r_2 - r_1), \quad q_0 = \max(r_1, r_2 - r_1), \tag{3.197}$$
$$\rho = n - r_2, \quad r = q + n - r_2, \quad p = q_0 - q. \tag{3.198}$$

We consider the differential equation

$$L_1 L_2 \cdots L_n u = 0, \quad x \in R^1, \quad t > 0. \tag{3.199}$$

Note that equation (3.199), as opposed to (3.1) for $n > r_2$ is not elliptic.

The equations of the form (3.199) for $r_2 > 0$ and $n > r_2$ are called equations of mixed type. Here we pose correct boundary value problems for the equations of mixed type (3.199).

The boundary conditions will be taken as

$$\frac{\partial^k u(x,0)}{\partial t^k} = f_k(x), \quad x \in R^1, \quad k = 0, \ldots, r-1, \tag{3.200}$$

$$Re\left[b_k(x)\frac{\partial^k u(x,0)}{\partial t^k}\right] = f_k(x), \quad x \in R^1, \quad k = r, \ldots, r+p-1, \tag{3.201}$$

where $f_0(x), \ldots, f_{r-1}(x)$ are given complex-valued and $f_r(x), \ldots, f_{r+p-1}(x)$ are real-valued functions satisfying $f_k(x) \in N_0(R^1)$ $(k = 0, \ldots, r+p-1)$. We suppose that the functions $b_k(x)$ $(k = q, \ldots, r+p-1)$ satisfy the same conditions as in (2.2) and the indices of $b_k(x)$ on X-axis are equal to zero.

A solution of the problem (3.199) – (3.201) is searched in the class $M_0(R^2_+)$.

In this section we prove the following

Theorem 3.8 *The problem (3.199) – (3.201) is uniquely solvable.*

To prove Theorem 3.8 we consider the following Cauchy problem

$$i\mu\frac{\partial u}{\partial x} + i\frac{\partial u}{\partial t} + au = h(x,t), \quad x \in R^1, \quad t > 0, \tag{3.202}$$

$$u(x,0) = f(x), \quad x \in R^1, \tag{3.203}$$

where μ is a real constant, $Im a \geq 0$, $h(x,t) \in M_0(R^2_+)$, and $f(x) \in N_0(R^1)$. A solution of this problem is searched in the class $M_0(R^2_+)$.

Lemma 3.4 *The Cauchy problem (3.202), (3.203) in the class $M_0(R^2_+)$ is uniquely solvable and the solution is defined by*

$$u(x,t) = \int_0^t e^{ia(t-\tau)} h(x - \mu t + \mu\tau, \tau) d\tau + e^{iat} f(x - \mu t). \tag{3.204}$$

First we find the general solution of equation (3.202) in the class of continuously differentiable functions. It is easy to check that

$$u_0(x,t) = \int_0^t e^{ia(t-\tau)} h(x - \mu t + \mu\tau, \tau) d\tau \tag{3.205}$$

is a particular solution of equation (3.202).

Consider the homogeneous equation corresponding to (3.202)

$$i\mu\frac{\partial v}{\partial x} + i\frac{\partial v}{\partial y} + av = 0. \tag{3.206}$$

Setting in (3.206) $v = e^{iat}w$ we get

$$\mu\frac{\partial w}{\partial x} + \frac{\partial w}{\partial y} = 0.$$

Changing the variables $\zeta = x - \mu t$ and $\eta = t$ we get

$$\frac{\partial w}{\partial \eta} = 0. \tag{3.207}$$

Hence $w = \Phi(\zeta)$, where $\Phi(\zeta)$ is an arbitrary continuously differentiable function on $(-\infty, \infty)$.

Returning to the variables x and t we get

$$v(x,t) = e^{iat}\Phi(x - \mu t). \tag{3.208}$$

From (3.205) and (3.208) we obtain that in the half-plane $x \in R^1$, $t > 0$ the general solution of equation (3.202) is defined by (3.204), with $\Phi(x-\mu t)$ instead of $f(x - \mu t)$. Thus from the boundary condition (3.203) we get (3.204).

We show now that this solution belongs to the class $M_0(R_+^2)$.

Let $t \geq 0,\ \tau \geq 0$. Then one has the inequality

$$\frac{1}{1+ \mid t - \tau \mid} \leq \frac{1+t}{1+\tau}.$$

Hence for $0 \leq \tau \leq t$, we have

$$\frac{1}{1+ \mid x - \mu t + \mu\tau \mid} \leq \frac{1}{1+ \mid \mid x \mid - \mid \mu \mid (t - \tau) \mid} \leq \frac{1 + 2 \mid \mu \mid t}{1+ \mid x \mid}. \tag{3.209}$$

Since $Ima \geq 0,\ f(x) \in N_0(R^1)$ and $h(x,t) \in M_0(R_+^2)$ then from (3.204) and (3.209) it follows that any solution of the problem (3.202), (3.203) belongs to the class $M_0(R_+^2)$.

Lemma 3.4 is proved.

Consider now the differential equation

$$L_{r_2+1} \cdots L_n u = h(x,t), \quad x \in R^1, \quad t > 0, \tag{3.210}$$

where $L_{r_2+1}, \ldots, L_n$ are the operators defined by (3.191), $h(x,t) \in M_0(R_+^2)$ and the constants μ_j and a_j $(j = r_2 + 1, \ldots, n)$ satisfy the conditions (3.194), (3.196).

As the boundary condition we take

$$\frac{\partial^k u(x,0)}{\partial t^k} = f_k(x), \quad k = 0,\dots,\rho-1, \quad x \in R^1, \tag{3.211}$$

where $f_k(x) \in N_0(R^2_+)$ $(k = 0,\dots,\rho-1)$.

We prove that the Cauchy problem (3.210), (3.211) in the class $M_0(R^2_+)$ is uniquely solvable.

Denote

$$L_{r_2+2}\cdots L_n u = v(x,t). \tag{3.212}$$

Equation (3.210) takes the form

$$L_{r_2+1} v = h(x,t). \tag{3.213}$$

From (3.212) it follows that

$$v(x,t) = \sum_{j+k\le\rho-1} a_{jk} \frac{\partial^{j+k} u(x,t)}{\partial x^j \partial t^k}, \tag{3.214}$$

where a_{jk} are some well-defined constants and

$$a_{0,\rho-1} = i^{\rho-1}. \tag{3.215}$$

In (3.214) and below we calculate the sum of all terms with non-negative integer indices j and k.

From (3.211) and (3.214) we obtain

$$v(x,0) = \sum_{j+k\le\rho-1} a_{kj} f_k^{(j)}(x). \tag{3.216}$$

From (3.214) and (3.215) it follows that for $k = \rho-1$ the condition (3.211) can be replaced by (3.216).

Thus, we have obtained the Cauchy problem (3.213), (3.216) to determine $v(x,t)$ which, according to Lemma 3.4, is uniquely solvable in the class $M_0(R^2_+)$. Substituting this solution into (3.212) we get the Cauchy problem for the equation (3.212) with boundary conditions (3.211) at $k = 0,\dots,\rho-2$.

Continuing as we have indicated above we conclude that the problem (3.210), (3.211) in the mentioned class is uniquely solvable. Here we also point out a method to solve the problem.

Now we turn to the investigation of the problem (3.199) – (3.201) and to the proof of Theorem 3.8. Along with (3.199) we consider the equations

$$L_{r_2+1}\cdots L_n u = v, \quad x \in R^1, \quad t > 0, \tag{3.217}$$

$$L_1\cdots L_{r_2} v = 0, \quad x \in R^1, \quad t > 0, \tag{3.218}$$

where $u(x,t)$ and $v(x,t)$ are solutions to be found, belonging to the class $M_0(R^2_+)$.

It is easy to check that if $u(x,t)$ is a solution of equation (3.199), then $u(x,t)$ and $v = L_{r_2+1}\cdots L_n u$ satisfy the system (3.217), (3.218) and vice versa, if $u(x,t)$ and $v(x,t)$ satisfy the system (3.217), (3.218), then $u(x,t)$ is a solution of equation (3.199). Hence, the problem (3.199) – (3.201) is equivalent to the problem (3.200), (3.201), (3.217), (3.218). Using some elementary transformations, we reduce the last problem to the above considered problems.

From (3.217) we have

$$v(x,t) = \sum_{j+k\le\rho} b_{jk}\frac{\partial^{j+k}u(x,t)}{\partial x^j\partial t^k}, \tag{3.219}$$

where b_{kj} are some well defined constants and $\rho = n - r_2$,

$$b_{0\rho} = (i)^\rho. \tag{3.220}$$

Differentiating both sides of (3.219) l times with respect to t and substituting $t = 0$, we get

$$\frac{\partial^l v(x,0)}{\partial t^l} = \sum_{j+k\le\rho} b_{jk}\Phi^{(j)}_{k+l}(x), \quad x\in R^1, \quad l = 0,\ldots,q_0-1, \tag{3.221}$$

where

$$\Phi_k(x) = \frac{\partial^k u(x,0)}{\partial t^k}, \quad k = 0,1,\ldots,n-1. \tag{3.222}$$

From (3.200) we have

$$\Phi_k(x) = f_k(x), \quad k = 0,\ldots,r-1, \quad x\in R^1. \tag{3.223}$$

Substituting $\Phi_k(x)$ from (3.223) into (3.221) we get

$$\frac{\partial^l v(x,0)}{\partial t^l} = g_l(x), \quad l = 0,\ldots,q-1, \tag{3.224}$$

$$\frac{\partial^l v(x,0)}{\partial t^l} = \sum_{j+k\le\rho} b_{kj}\Phi^{(j)}_{k+l}(x), \quad l = q,\ldots,q_0-1, \tag{3.225}$$

where

$$g_l(x) = \sum_{j+k\le\rho} b_{jk}f^{(j)}_{k+l}(x), \quad l = 0,\ldots,q-1. \tag{3.226}$$

From (3.220) and (3.223) it follows that the system (3.225) is uniquely solvable with respect to $\Phi_\nu(x)$ ($\nu = r,\ldots,r+p-1$) and the solution has the form

$$\frac{\partial^\nu u(x,0)}{\partial t^\nu} \equiv \Phi_\nu(x) = \sum_{j+k\le\nu-\rho} A_{\nu jk}\frac{\partial^{j+k}v(x,0)}{\partial x^j\partial t^k} + \psi_\nu(x), \tag{3.227}$$

where $A_{\nu jk}$ are some well-defined constants and

$$A_{\nu jk} = \begin{cases} (-i)^k, & \text{for } j=0,\ k=\nu-\rho,\ \nu=r,\ldots,r+p-1, \\ 0, & \text{for } \ k<q \end{cases}$$

and $\psi_\nu(x)$ are linear expressions on $f_k(x)$ $(k = 0,\ldots,r-1)$ and their derivatives of order up to $n-1$.

Substituting the values of $\frac{\partial^\nu u(x,0)}{\partial t^\nu}$ from (3.227) into the boundary condition (3.201) we get

$$Re \sum_{j+k\leq l} b_{\rho+l}(x) B_{ljk} \frac{\partial^{j+k} v(x,0)}{\partial x^j \partial t^k} = \omega_l(x), \quad l = q,\ldots,q_0-1, \tag{3.228}$$

where B_{ljk} are well-defined constants and

$$B_{lkj} \neq 0 \quad \text{for} \quad j=0, \quad k=l, \quad l=q,\ldots,q_0-1, \tag{3.229}$$

$$B_{ljk} = 0 \quad \text{for} \quad k<q,$$

and $\omega_l(x)$ $(l = q,\ldots,q_0-1)$ are linear expressions on $f_j(x)$ $(j = 1,\ldots,n-1)$ and their derivatives of order up to $n-1$. So we can replace the boundary conditions (3.201) by the boundary conditions (3.228).

Take the first ρ conditions from (3.200):

$$\frac{\partial^k u(x,0)}{\partial t^k} = f_k(x), \quad x \in R^1, \quad k = 0,\ldots,\rho-1. \tag{3.230}$$

Let $u(x,t)$ and $v(x,t)$ be the solutions of the system (3.217) and (3.218). Then it is obvious that the conditions (3.224), (3.230) follow from (3.200). The converse is also true: if the functions $u(x,t)$ and $v(x,t)$ satisfy the system (3.217), (3.218), then the condition (3.200) follows from the conditions (3.224) and (3.230). Therefore the conditions (3.200) and (3.201) can be replaced by the conditions (3.224), (3.228) and (3.230).

The problem (3.199) – (3.201) can be resolved in the following way. First, we resolve the problem (3.218), (3.224), (3.228) with respect to unknown function $v(x,t)$, and then we resolve the problem (3.217), (3.230) with respect to unknown function $u(x,t)$. As we have shown in this chapter both problems are uniquely solvable in the class $M_0(R^2_+)$.

Thus, the problem (3.199) – (3.201) is uniquely solvable.

Theorem 3.8 is proved.

3.12 Normal Solvability of Dirichlet-Type Problem for Products of First Order Differential Operators

Let L_j be the differential operators defined by (3.191) with constants μ_j and a_j ($j = 1, \ldots, n$) satisfying the conditions (3.192) – (3.196).

Consider the equation

$$L_1 L_2 \cdots L_n u = 0, \quad x \in R^1, \quad t > 0 \tag{3.231}$$

with the boundary conditions

$$\frac{\partial^k u(x,0)}{\partial t^k} = f_k(x), \quad x \in R^1, \quad k = 0, \ldots, r-1, \tag{3.232}$$

$$Re \sum_{j+l \leq k} b_{kjl} \frac{\partial^{j+l} u(x,0)}{\partial x^j \partial t^l} = f_k(x), \ x \in R^1, \ k = r, \ldots, r+p-1, \tag{3.233}$$

where $f_0(x), \ldots, f_{r-1}(x)$ are given complex-valued and $f_r(x), \ldots, f_{r+p-1}(x)$ are real-valued functions, $f_k(x) \in N_0(R^1)$ ($k = 0, \ldots, r+p-1$) and b_{kjl} are constants. These constants can be complex or real. The integers r and p are defined by (3.197), (3.198). A solution of the problem (3.231) – (3.233) is searched in the class $M_0(R_+^2)$.

Definition 3.2. We will say that the problem (3.231) – (3.233) is normally solvable if the non-homogeneous problem (3.231) – (3.233) is always solvable and the corresponding homogeneous problem (at $f_j \equiv 0$, $j = 0, \ldots, r+p-1$) has only a finite number of linear independent solutions.

One has the following

Theorem 3.9 *Assume that the coefficients of (3.233) satisfy*

$$b_{k0k} \neq 0, \quad k = r, \ldots, r+p-1, \tag{3.234}$$

then the problem (3.231) – (3.233) is uniquely solvable and if for any non-negative integer ρ ($r \leq \rho \leq r+p-1$) the equality $b_{\rho 0 \rho} = 0$ holds, then the homogeneous problem, corresponding to the problem (3.231) – (3.233), has an infinite number of linearly independent solutions.

From Theorem 3.9 we immediately deduce

Corollary 3.1 *The condition (3.234) is necessary and sufficient for normal solvability of the problem (3.231) – (3.233).*

Proof of Theorem 3.9. Let the condition (3.234) be satisfied. Then as in the case of the problem (3.199) – (3.201) we can show that the problem (3.229) – (3.231) is also uniquely solvable.

Now for some non-negative integer ν ($r \leq \nu \leq r+\rho-1$), let the equality

$$b_{\nu 0 \nu} = 0 \tag{3.235}$$

be satisfied. We prove that in this case the homogeneous problem (3.231) – (3.233) in the class $M_0(R^2_+)$ has infinitely many linear independent solutions.

To this end we consider two cases.

Case 1. Let $r_2 = n$. This means that equation (3.231) is elliptic. For the sake of definiteness, we assume that $r_1 \geq r_2 - r_1$.

Without loss of generality, we can assume that

$$b_{kk} \neq 0 \quad for \quad k = \nu+1, \ldots, r+p-1. \tag{3.236}$$

This is always possible if we choose ν as the maximal of indices satisfying (3.235).

Consider the system of equations

$$\sum_{j+l \leq k} b_{kjl}\psi_l^{(j)}(z) = 0, \quad k = \nu+1, \ldots, r+p-1, \quad Imz > 0, \tag{3.237}$$

where $\psi_\nu(z)$ is a given function on complex variable $z = x+iy$ which is analytic in $Imz > 0$ and $\psi_{\nu+1}(z), \ldots, \psi_{r+p-1}(z)$ are analytic functions to be found. We suppose that $\psi_\nu(z) \in N_0(R^2_+)$.

From the system (3.237) at $k = \nu+1$ we find $\psi_{\nu+1}(z)$. Substituting $\psi_{\nu+1}(z)$ into (3.237) at $k = \nu+2$ we find $\psi_{\nu+2}(z)$ and so on.
Thus, the system (3.237) is uniquely solvable. It is easy to see that a solution of this system also belongs to the class $N_0(R^2_+)$.

Consider now the following Cauchy problem

$$L_1 \cdots L_{r_1} u = 0, \quad x \in R^1, \quad t > 0, \tag{3.238}$$

$$\frac{\partial^k u(x,0)}{\partial t^k} = 0, \quad k = 0, \ldots, \nu-1, \tag{3.239}$$

$$\frac{\partial^k u(x,0)}{\partial t^k} = \psi_j(x), \quad j = \nu, \ldots, r+p-1, \tag{3.240}$$

where $\psi_\nu(z)$ is a given analytic in R^2_+ function, $\psi_\nu(z) \in N_0(R^2_+)$ and $\psi_{\nu+1}(z), \ldots, \psi_{r+p-1}(z)$ satisfy the system (3.237).

Let us show that a solution of the problem (3.238) – (3.240) satisfy the homogeneous problem (3.231) – (3.233).

Indeed, if the function $u(x,t)$ is a solution of equation (3.238) then it is evident that it is also a solution of equation (3.231) (because $L_jL_k = L_kL_j$). Substituting the values of $\frac{\partial^k u(x,0)}{\partial t^k}$ from (3.239), (3.240) into (3.232) and (3.233), and taking into account (3.237) we conclude that $u(x, y)$ satisfies the conditions (3.232), (3.233) at $f_k \equiv 0 \quad (k = 0, \ldots, r + p - 1)$.

In section 3.3 we have shown that the Cauchy problem (3.238) – (3.240) in the class $M_0(R^2_+)$ is always uniquely solvable. Let $u_n(x)$ be a solution of the Cauchy problem (3.238) – (3.240) with

$$\psi_\nu(x) = (x + i)^{-n},$$

where n is an integer. Since the functions $(x + i)^{-1}, (x + i)^{-2}, \ldots$ are linearly independent, then $u_1(z), \ldots, u_n(z), \ldots$ also are linearly independent.

Thus, we have constructed infinitely many linearly independent solutions for the homogeneous problem (3.231) – (3.233).

Hence in this case the second part of Theorem 3.9 is proved.

Case 2. Let $r_2 < n$. In section 3.11 we have reduced the problem (3.199) – (3.201) to the equivalent problem (3.217), (3.218), (3.224), (3.228), (3.230). In a similar way the homogeneous problem (3.231) – (3.233) (at $f_k(x) \equiv 0, k = 0, \ldots, r + p - 1$) can be reduced to the problem

$$L_1 \cdots L_{r_2} v = 0, \quad t > 0, \quad x \in R^1, \tag{3.241}$$

$$\frac{\partial^l v(x, 0)}{\partial t^l} = 0, \quad l = 0, \ldots, q - 1, \quad x \in R^1, \tag{3.242}$$

$$Re \sum_{j+l \leq k} B_{kjl} \frac{\partial^{j+k} v(x, 0)}{\partial x^j \partial t^l} = 0, \quad l = q, \ldots, q_0 - 1, \quad x \in R^1, \tag{3.243}$$

$$L_{r_2+1} \cdots L_n u = v(x, t), \quad t > 0, \quad x \in R^1, \tag{3.244}$$

$$\frac{\partial^k u(x, 0)}{\partial t^k} = 0, \quad k = 0, \ldots, n - r_2 - 1, \quad x \in R^1, \tag{3.245}$$

where B_{kjl} are some constants such that $B_{\nu 0 \nu} = 0$ provided that $b_{\nu 0 \nu} = 0$.

Equation (3.241) is elliptic. Hence from the condition $B_{\nu 0 \nu} = 0$ it follows that the problem (3.241) – (3.243) in the class $M_0(R^2_+)$ has infinitely many linearly independent solutions $v_1(x, t), \ldots, v_n(x, t), \ldots$, belonging to the class $M_0(R^2_+)$.

Let $u_n(x, t)$ be a solution of the problem (3.244), (3.245) at $v = v_n(x, t)$. Since $v(x, t) \in M_0(R^2_+)$ then, as we have proved in section 3.11 (see problem (3.210), (3.211)) the solution $u_n(x, t)$ belongs to the class $M_0(R^2_+)$. It is clear that the functions $u_1(x, t), \ldots, u_n(x, t), \ldots$ are also linearly independent. Hence, in this case the homogeneous problem (3.231) - (3.233) has infinitely many linearly independent solutions.

Theorem 3.9 is proved.

In particular, it follows from this result that in (3.201) the assumption $b_k(x) \neq 0, (k = r, \ldots, r + p - 1)$ is natural.

3.13 A Method to Solve Cauchy Problem for Elliptic and Hyperbolic Equations

Consider the equation

$$\sum_{k=1}^{n} A_k \frac{\partial^n u}{\partial x^k \partial t^{n-k}} = 0, \quad x \in R^1, \quad t > 0, \tag{3.246}$$

where $A_0, \ldots, A_n$ are some constants, $A_0 \neq 0$ and the roots $\lambda_1, \ldots, \lambda_n$ of the equation

$$A_0 \lambda^n + A_1 \lambda^{n-1} + \cdots + A_n = 0, \tag{3.247}$$

satisfy the condition $Im\lambda_k \neq 0, \ k = 1, \ldots, n$.

Here the roots are counted with their multiplicities. Observe that $Im\lambda_k \neq 0 \ (k = 1, \ldots, n)$ are conditions of ellipticity of equation (3.246).

Equation (3.246) can be represented as

$$L_1 L_2 \cdots L_n u = 0, \tag{3.248}$$

where

$$L_k u \equiv \frac{\partial u}{\partial t} - \lambda_k \frac{\partial u}{\partial x}.$$

Hence equation (3.246) is a special case of equation (3.2), where $a_j = 0 \ (j = 1, \ldots, n)$.

Let r be the number of roots of equation (3.247) with $Im\lambda > 0$ and $q = n-r$ the number of roots of this equation with $Im\lambda < 0$.

For the sake of definiteness, assume that $r \geq q$. As the boundary conditions for equation (3.246) we take (3.6) and (3.7). According to Theorem 3.1, this equation under such conditions is uniquely solvable. In section 3.5 the resolution of this problem have been reduced to that of Cauchy problem for the equation

$$L_1 L_2 \cdots L_r u = 0 \tag{3.249}$$

and

$$L_{r+1}^* \cdots L_n^* v = 0, \tag{3.250}$$

where

$$L_k^* v = \frac{\partial v}{\partial t} - \overline{\lambda}_k \frac{\partial v}{\partial x}.$$

As the boundary conditions for Cauchy problem for (3.249) we take (cf. section 3.3)

$$\frac{\partial^\nu u(x,0)}{\partial t^\nu} = \psi_\nu(x), \quad \nu = 0, \ldots, r-1, \tag{3.251}$$

where $\psi_\nu(z)$ are given analytic in R^2_+ functions belonging to the class $N_0(R^2_+)$.

In section 3.3 it has been shown that the Cauchy problem (3.249), (3.251) is uniquely solvable and the solution belongs to the class $M_0(R^2_+)$ (see Theorem 3.2). In this section we propose another more efficient method of resolution of this problem. First, we find the general solution of equation (3.249) in the half-plane $Imz > 0$.

As we have shown in chapter 2, the general solution of equation

$$\frac{\partial u}{\partial t} - \lambda \frac{\partial u}{\partial x} = 0, \quad t > 0, \quad x \in R^1, \quad Im\lambda > 0, \tag{3.252}$$

is defined by

$$u(x,t) = \varphi(x + \lambda t), \tag{3.253}$$

where $\varphi(z)$ is arbitrary analytic in R^2_+ function. Since $Im\lambda > 0$ and $t > 0$, then $Im(x + \lambda t) > 0$.

Let us consider equation

$$\frac{\partial u}{\partial t} - \lambda \frac{\partial u}{\partial x} = t^k \varphi(x + \lambda t), \quad t > 0, \quad x \in R^1, \tag{3.254}$$

where $\varphi(x + \lambda t)$ is the function from (3.253).

It is easy to check that the function

$$u(x,t) = \frac{t^{k+1}}{k+1} \varphi(x + \lambda t) \tag{3.255}$$

is a particular solution of equation (3.254).

Hence, the general solution of equation (3.254) is defined by

$$u(x,t) = \frac{t^{k+1}}{k+1} \varphi(x + \lambda t) + \psi(x + \lambda t), \tag{3.256}$$

where $\psi(z)$ is arbitrary analytic in R^2_+ function.

Consider now the equation

$$\frac{\partial u}{\partial t} - \mu \frac{\partial u}{\partial x} = t^k \varphi(x + \lambda t), \tag{3.257}$$

where $\lambda \neq \mu$, $Im\mu > 0$.

The solution is searched as

$$u(x,t) = \sum_{j=0}^{k} \varphi_j(x + \lambda t) t^j, \tag{3.258}$$

where $\varphi_j(x+\lambda t)$ are unknown functions, which are analytic on $x+\lambda t$. Substituting $u(x,t)$ from (3.258) into (3.257) we get

$$\sum_{j=0}^{k}[\varphi_j'(x+\lambda t)(\lambda-\mu)t^j + jt^{j-1}\varphi_j(x+\lambda t)] = t^k\varphi(x+\lambda t). \tag{3.259}$$

Equating the coefficients of t^j $(j = 0,\ldots,k)$, we get

$$\varphi_k'(z)(\lambda-\mu) = \varphi(z), \quad z = x+it, \quad Imz > 0, \tag{3.260}$$
$$\varphi_j'(z)(\lambda-\mu) + (j+1)\varphi_{j+1}(z) = 0, \; Imz > 0, \; j = 0,1,\ldots,k-1. \tag{3.261}$$

Resolving the system (3.260), (3.261) we find the analytic functions $\varphi_0(z),\ldots,$ $\varphi_k(z)$.

Hence the general solution of equation (3.257) is defined by

$$u(x,t) = \sum_{j=0}^{k}\varphi_k(x+\lambda t)t^j + \psi(x+\mu t), \tag{3.262}$$

where $\varphi_0(z),\ldots,\varphi_k(z)$ is a particular solution of the system (3.260), (3.261) and $\psi(z)$ is arbitrary analytic in R^2_+ function.

Now let $\mu_1,\ldots,\mu_\rho$ be distinct roots of equation (3.247) with $Re\lambda > 0$ and let $k_1,\ldots,k_\rho$ be their multiplicities, $k_1+\cdots+k_\rho = r$.

We find now the general solution of equation (3.249). Denote

$$v_k(x,t) = L_{k+1}\cdots L_r u, \quad k = 1,2,\ldots,r-1. \tag{3.263}$$

It is clear that

$$L_k v_k(x,t) = v_{k-1}(x,t), \quad k = 2,\ldots,r-1, \tag{3.264}$$
$$L_r u = v_{r-1}. \tag{3.265}$$

From (3.263) we have $v_1 = L_2\cdots L_r u$. Hence equation (3.249) can be represented as

$$L_1 v_1 = 0. \tag{3.266}$$

Thus, if u is a solution of equation (3.249) then $u, v_1,\ldots,v_{r-1}$ is a solution of the system (3.264), (3.265), (3.266). The converse is also true, if the functions $u, v_1,\ldots,v_{r-1}$ satisfy the system of equations (3.264), (3.265), (3.266), then $u(x,t)$ is a solution of equation (3.249).

The general solution of the system (3.264), (3.265), (3.266) can be obtained in the following way.

First we find the general solution of equation (3.266). Substituting this solution into (3.264) at $k = 2$, we get the general solution of equation $L_2 v_2 =$

$v_1(x,t)$. Continuing in the same manner we get the general solution of the system (3.264), (3.265), (3.266). In fact, each time we resolve an equation of the form (3.254) or (3.257).

Therefore, any solution of equation (3.249) has the form

$$u(x,t) = \sum_{j=1}^{\rho} \sum_{l=0}^{k_j-1} \varphi_{jl}(x+\mu_j t)t^l, \tag{3.267}$$

where $\varphi_{jl}(x+\mu_j t)$ $(l = 0,\ldots,k_j-1)$ are some analytic on $x+\mu_j t$ in $Imz > 0$ functions.

It is clear that any analytic function $\varphi_{jl}(x+\mu_j t)$ can be represented as

$$\varphi_{jl}(x+\mu_j t) = \Phi_{jl}^{(l)}(x+\mu_j t), \tag{3.268}$$

where $\Phi_{jl}^{(l)}(z)$ is the l-th order derivative of analytic function $\Phi_{jl}(z)$.

Substituting $\varphi_{jl}(x+\mu_j t)$ from (3.268) into (3.267) we get

$$u(x,t) = \sum_{j=1}^{\rho} \sum_{l=0}^{k_j-1} \Phi_{jl}^{(l)}(x+\mu_j t)t^l. \tag{3.269}$$

On the other hand, it is easy to check that any function (3.269) satisfies equation (3.249). Hence the general solution of equation (3.249) is defined by (3.269), where $\Phi_{jl}(z)$ is arbitrary analytic in R^2_+ function.

Substituting the general solution (3.269) into the condition (3.251) we get

$$\sum_{j=1}^{\rho} \sum_{l=0}^{k_j-1} A_{\nu jl}\Phi_{jl}^{(\nu)}(x) = \psi_\nu(x), \quad \nu = 0,\ldots,r-1, \quad x \in R^1, \tag{3.270}$$

where $A_{\nu jl}$ are some constants.

Since $\Phi_{jl}(z)$ and $\psi_\nu(z)$ are analytic on $z = x+iy$ in $Imz > 0$, it follows from (3.270) that

$$\sum_{j=1}^{\rho} \sum_{l=0}^{k_j-1} A_{\nu jl}\Phi_{jl}^{(\nu)}(z) = \psi_\nu(z), \quad \nu = 0,\ldots,r-1, \quad Imz > 0. \tag{3.271}$$

Integrating (3.270) ν $(\nu = 0,\ldots,r-1)$ times we get

$$\sum_{j=1}^{\rho} \sum_{l=0}^{k_j-1} A_{\nu jl}\Phi_{jl}(z) = \omega_\nu(z) + P_{\nu-1}(z), \quad Imz > 0, \tag{3.272}$$

where

$$\omega_\nu(z) = \frac{1}{(\nu-1)!}\int_0^z (z-\zeta)^{\nu-1}\psi_\nu(\zeta)d\zeta, \tag{3.273}$$

and $P_{\nu-1}(z)$ is an arbitrary polynomial of degree at most $\nu - 1$. Without loss of generality, we take $P_{\nu-1}(z) \equiv 0$.

We are going to prove that the system (3.272) is uniquely solvable with respect to $\Phi_{jl}(z)$ ($l = 0,\dots,k_j - 1$; $j = 1,\dots,\rho$). Resolving the system (3.272) with $P_{\nu-1} \equiv 0$ and substituting (3.269) into the general solution, we find a solution of the Cauchy problem (3.249), (3.251). Since $\psi_\nu(z) \in N_0(R^2_+)$ then, as we have mentioned above, this solution belongs to the class $M_0(R^2_+)$ (see section 3.3).

To complete our investigation of the problem (3.249), (3.251) we prove the following result.

Lemma 3.5 *Equation (3.272) is uniquely solvable with respect to functions* $\Phi_{jl}(z)$ ($l = 0,\dots,k_j - 1$; $j = 1,\dots,\rho$).

Proof. First we prove that the homogeneous system (3.272) (at $\omega_\nu \equiv 0$, $P_{\nu-1} \equiv 0$) in the class $N_0(R^2_+)$ has only zero solution.

Let $\Phi_{jl}(z)$ ($l = 0,\dots,k_j - 1$; $j = 1,\dots,\nu$) be a solution of the homogeneous system (3.272) belonging to the class $N_0(R^2_+)$. Substituting this solution into the general solution (3.269) we get a solution of the Cauchy homogeneous problem (3.249), (3.251) (at $\psi_\nu \equiv 0$). Since the problem (3.249), (3.251) is uniquely solvable, the solution is equal to zero, i.e.

$$u(x,y) \equiv \sum_{j=1}^{\rho}\sum_{l=0}^{k_j-1} \Phi^{(l)}_{jl}(x+\mu_j t)t^l \equiv 0, \quad x \in R^1, \quad t > 0. \tag{3.274}$$

Applying the operator $L_1^{k_1-1}L_2^{k_2}\cdots L_\rho^{k_\rho}$ to both sides of (3.272) we get

$$\Phi^{(l)}_{1,k_1-1}(z) \equiv 0 \quad \text{for} \quad l = k_1 + k_2 + \cdots + k_\rho - 1. \tag{3.275}$$

Since $\Phi^{(l)}_{1,k_1-1}(z) \in N_0(R^2_+)$, it follows from (3.275) that $\Phi_{1,k_1-1}(z) \equiv 0$. Continuing this process we get

$$\Phi_{jl}(z) = 0, \quad l = 0,\dots,k_j - 1, \quad j = 1,\dots,\rho.$$

Hence the homogeneous equation (3.272) in the class $N_0(R^2_+)$ has only zero solution. From here it immediately follows that equation (3.272) is uniquely solvable.

Lemma 3.5 is proved.

Consider now the equation

$$\sum_{k=0}^{r} B_k \frac{\partial^r u}{\partial x^k \partial t^{r-k}} = 0, \quad x \in R^1, \quad t > 0. \tag{3.276}$$

Let the equation

$$B_0 \lambda^r + B_1 \lambda^{r-1} + \cdots + B_r = 0.$$

have only real roots. The solution is searched in the class of functions having continuous derivatives up to order r. Let $\mu_1, \ldots, \mu_\rho$ be distinct roots of this equation with multiplicities $k_1, \ldots, k_\rho$ $(k_1 + \cdots + k_\rho = r)$.

Using the above arguments, we can prove that the general solution of equation (3.276) is defined by (3.269), where in this case $\Phi_j(x)$ are arbitrary functions having continuous derivatives up to order r. Using the general solution of equation (3.276) we resolve the Cauchy problem for this equation. As the boundary conditions for Cauchy problem for (3.276) we take (cf. section 3.3)

$$\frac{\partial^\nu u(x,0)}{\partial t} = \psi_\nu(x), \quad \nu = 0, \ldots, r-1, \tag{3.277}$$

where $\psi_j(x)$ is a given function in R^1 and $\psi_\nu(x) \in N_0(R^1)$.

Chapter 4

Propagation of Plane Periodic Electromagnetic Waves in Stratified Medium

4.1 Introduction

Let us write the complete system of equations for electromagnetic theory, i.e. the well-known partial differential equations of Maxwell (see [18], p. 27 and [4], p. 49)

$$rot\mathbf{H} = \frac{\partial \mathbf{D}}{\partial t} + \mathbf{j}, \quad div\mathbf{D} = \rho, \tag{4.1}$$

$$rot\mathbf{E} = -\frac{\partial \mathbf{B}}{\partial t}, \quad div\mathbf{B} = 0, \tag{4.2}$$

and the algebraic relations

$$\mathbf{D} = \varepsilon\varepsilon_0\mathbf{E}, \quad \mathbf{B} = \mu\mu_0\mathbf{H}, \quad \mathbf{j} = \sigma\mathbf{E}, \tag{4.3}$$

where $\mathbf{E} = \mathbf{E}(\vec{r}, t)$ is the intensity of an electric field, $\vec{r}$ is a radius-vector of a point (x, y, z), t is time, $\mathbf{H} = \mathbf{H}(\vec{r}, t)$ is the intensity of a magnetic field, $\mathbf{D} = \mathbf{D}(\vec{r}, t)$ is an electric induction, $\mathbf{B} = \mathbf{B}(\vec{r}, t)$ is a magnetic induction, ε_0 is the electric constant, $\varepsilon_0 = 10^7/4\pi C^2 \approx 8,854 \cdot 10^{-7}$, C is the speed of light in vacuum, μ_0 is the magnetic constant $\mu_0 = 4\pi \cdot 10^{-7} \approx 1,257 \cdot 10^{-6}$, $\mathbf{j}$ is a conductance current density, ρ is a charge density, ε is a dielectric constant, μ is a magnetic permeability and σ is a specific electric conductivity.

The values of **H**, **D**, **j**, **E** and **B** are vectors in three-dimensional space with Cartesian coordinates $(OXYZ)$, where O is the origin.

Taking into account (4.3), the system (4.1), (4.2) can be written as

$$\rho = \varepsilon_0 div(\varepsilon \mathbf{E}), \tag{4.4}$$

$$\mu_0 \frac{\partial(\mu \mathbf{H})}{\partial t} = -rot\mathbf{E}, \quad div(\mu\mathbf{H}) = 0, \tag{4.5}$$

$$\varepsilon_0 \frac{\partial(\varepsilon \mathbf{E})}{\partial t} = rot\mathbf{H} - \sigma\mathbf{E}. \tag{4.6}$$

Thus, the investigation of the system (4.1) – (4.3) is reduced to that of the system (4.4) – (4.6).

In this chapter we consider electromagnetic fields, depending only on x and time t, which are periodic on t with period T. Without loss of generality we assume that $T = 2\pi$. The medium in which the electromagnetic field will be considered, consists of one, two or more homogeneous strata. The parameters ε, μ and σ are scalar and have different values in different strata, i.e. they take piecewise constant values.

The system of equations (4.4) – (4.6) will be considered in the domain

$$R^2_+ = \{(x,t); x > 0, t \in (-\infty, +\infty)\}.$$

Based on the physical considerations, a solution of the system (4.4) – (4.6) will be sought in the class of bounded functions.

At the end of the book we also consider harmonic oscillations of electromagnetic waves in non-homogeneous medium.

4.2 Boundary Value Problem for the System (4.4) – (4.6) in Homogeneous Medium

Let the medium consist of one homogeneous isotropic conducting stratum $x > 0$. This means that the parameters ε, μ and σ are positive constants. Denote by E_x, E_y, E_z and H_x, H_y, H_z the components of intensities of electric **E** and magnetic **H** fields. In the case, where the electromagnetic field depends only on x and t, the system of equations (4.4) – (4.6) takes the form

$$\rho(x,t) = \varepsilon_0 \frac{\partial(\varepsilon E_x(x,t))}{\partial x}; \tag{4.7}$$

$$
\begin{aligned}
\varepsilon_0 \frac{\partial(\varepsilon E_x(x,t))}{\partial t} &= -\sigma E_x(x,t), \\
\frac{\partial(\mu H_x(x,t))}{\partial t} &= 0, \\
\frac{\partial(\mu H_x(x,t))}{\partial x} &= 0;
\end{aligned}
\tag{4.8}
$$

$$
\begin{aligned}
\varepsilon_0 \frac{\partial(\varepsilon E_y(x,t))}{\partial t} &= -\frac{\partial H_z(x,t)}{\partial x} - \sigma E_y(x,t), \\
\mu_0 \frac{\partial(\mu H_z(x,t))}{\partial t} &= -\frac{\partial E_y(x,t)}{\partial x};
\end{aligned}
\tag{4.9}
$$

$$
\begin{aligned}
\mu_0 \frac{\partial(\mu H_y(x,t))}{\partial t} &= \frac{\partial E_z(x,t)}{\partial x}, \\
\varepsilon_0 \frac{\partial(\varepsilon E_z(x,t))}{\partial t} &= \frac{\partial H_y(x,t)}{\partial x} - \sigma E_z(x,t).
\end{aligned}
\tag{4.10}
$$

A periodic on t solution of the system (4.8) is defined by

$$
\mu H_x(x,t) \equiv const, \quad E_x(x,t) \equiv 0, \quad (x,t) \in R^2_+. \tag{4.11}
$$

Substituting $E_x(x,t)$ from (4.11) into (4.7) we get

$$
\rho(x,t) \equiv 0, \quad (x,t) \in R^2_+. \tag{4.12}
$$

The systems (4.9) and (4.10) have an infinite number of linearly independent solutions, belonging to the above mentioned class.

In this section we mention some boundary conditions, determining the solutions of these systems uniquely or up to a finite number of linearly independent solutions.

Note that the corresponding boundary value problem will only be investigated for the system (4.10). The system (4.9) can be reduced to (4.10) by substitution

$$
H_z(x,t) = -u(x,t), \quad E_y(x,t) = v(x,t).
$$

By assumption the components $H_y(x,t)$ and $E_z(x,t)$ of solutions of the system (4.10) are bounded in R^2_+ and are 2π-periodic on t, i.e.

$$
\begin{aligned}
&| H_y(x,t) | \leq const, \quad | E_z(x,t) | \leq const, \quad (x,t) \in R^2_+, \\
&H_y(x,t+2\pi) = H_y(x,t), \quad E_z(x,t+2\pi) = E_z(x,t), \quad (x,t) \in R^2_+.
\end{aligned}
$$

We seek a solution of the system (4.10) in the class of infinitely differentiable in R^2_+ and continuous in $\overline{R^2_+}$, functions, where

$$\overline{R^2_+} = \{(x,t), \quad x \geq 0, \quad t \in (-\infty, +\infty)\}.$$

The boundary condition for the system (4.10) will be taken as

$$aH_y(0,t) + bE_z(0,t) = f(t). \tag{4.13}$$

Here a and b are real constants, $a^2 + b^2 \neq 0$, $f(t)$ is a given infinitely differentiable 2π-periodic function.

The problem (4.10), (4.13) at $f \equiv 0$ is said to be homogeneous.

One has the following theorems:

Theorem 4.1 *If $a \neq 0$, then the problem (4.10), (4.13) is uniquely solvable.*

Theorem 4.2 *If $a = 0$, then the homogeneous problem (4.10), (4.13) has only one linear independent solution $H_y(x,t) \equiv const$, $E_z(x,t) \equiv 0$ and the corresponding non-homogeneous problem is solvable if and only if*

$$\int_0^{2\pi} f(t)dt = 0. \tag{4.14}$$

Proof of Theorem 4.1. Let $a \neq 0$. Without loss of generality, we can assume that $a = 1$, i.e. the boundary condition (4.13) has the form

$$H_y(0,t) + bE_z(0,t) = f(t). \tag{4.15}$$

Let $B_y(x,t)$ and $E_z(x,t)$ be infinitely differentiable on x and t and 2π-periodic on t real solutions of the system (4.10). This solution admits the following Fourier series expansion:

$$H_y(x,t) = u_0(x) + 2Re\sum_{n=1}^{\infty} u_n(x)e^{int}, \quad x > 0, \quad t \in (-\infty, +\infty), \tag{4.16}$$

$$E_z(x,t) = v_0(x) + 2Re\sum_{n=1}^{\infty} v_n(x)e^{int}, \quad x > 0, \quad t \in (-\infty, +\infty), \tag{4.17}$$

where i is the imaginary unit, $u_n(x)$ and $v_n(x)$ $(n = 0, 1, \ldots)$ are the Fourier coefficients of these solutions:

$$u_n(x) = \frac{1}{2\pi}\int_0^{2\pi} H_y(x,\tau)e^{-in\tau}d\tau, \quad v_n(x) = \frac{1}{2\pi}\int_0^{2\pi} E_z(x,\tau)e^{-in\tau}d\tau, \tag{4.18}$$

where $n = 0, 1, \ldots$.

It follows from (4.18) that the functions $u_n(x)$ and $v_n(x)$ ($n = 0, 1, \ldots$) are bounded on semi-axis $x \geq 0$ and $u_0(x)$ and $v_0(x)$ are real-valued.

Since $H_y(x,t)$ and $E_z(x,t)$ satisfy the system (4.10), it follows from (4.18) that $u_n(x)$ and $v_n(x)$ ($n = 0, 1, \ldots$) satisfy the following system of ordinary differential equations

$$\frac{du_n(x)}{dx} = (in\varepsilon\varepsilon_0 + \sigma)v_n(x), \quad \frac{dv_n(x)}{dx} = in\mu\mu_0 u_n(x), \quad x > 0. \tag{4.19}$$

The system (4.19) can also be obtained by substituting $H_y(x,t)$ and $E_z(x,t)$ from (4.16) and (4.17) into the system (4.10).

From (4.19), excluding $u_n(x)$ for $n \neq 0$, we get

$$u_n(x) = \frac{1}{in\mu\mu_0} \cdot \frac{dv_n(x)}{dx}, \tag{4.20}$$

$$\frac{d^2v_n(x)}{dx^2} = n\mu\mu_0(-n\varepsilon\varepsilon_0 + i\sigma)v_n(x), \quad x > 0. \tag{4.21}$$

Observe that

$$\lambda^2 = n\mu\mu_0(-n\varepsilon\varepsilon_0 + i\sigma)$$

is the characteristic equation of (4.21), the roots of which are $\pm\lambda_n$, where

$$\lambda_n = -in\sqrt{\alpha}\sqrt{1 - \frac{i\beta}{n}} = -a_n - ib_n, \tag{4.22}$$

with

$$a_n = \frac{\beta\sqrt{\alpha}}{\sqrt{2}\sqrt{1 + \sqrt{1 + \beta^2 n^{-2}}}}, \quad b_n = \frac{n\sqrt{\alpha}}{\sqrt{2}}\sqrt{1 + \sqrt{1 + \beta^2 n^{-2}}}$$

and $\alpha = \varepsilon\varepsilon_0\mu\mu_0$, $\beta = \sigma(\varepsilon\varepsilon_0)^{-1}$.

Hence, the general solution of equation (4.21) belonging to the class of functions bounded on $x \geq 0$ is defined by

$$v_n(x) = c_n e^{\lambda_n x}, \quad x > 0, \quad n = 1, 2, \ldots, \tag{4.23}$$

where c_n is an arbitrary complex constant.

Substituting $v_n(x)$ from (4.23) into (4.20) we get

$$u_n(x) = \frac{c_n\lambda_n}{in\mu\mu_0}e^{\lambda_n x}, \quad x > 0, \quad n = 1, 2, \ldots. \tag{4.24}$$

Let $n = 0$. Then the bounded solution of the system (4.19) is defined by

$$u_0(x) = c_0, \quad v_0(x) = 0, \quad x > 0, \tag{4.25}$$

where c_0 is a real constant.

From (4.18) we have

$$u_n(0) + bv_n(0) = \frac{1}{2\pi}\int_0^{2\pi}[H_y(0,\tau) + bE_z(0,\tau)]e^{-in\tau}d\tau. \tag{4.26}$$

From here and the boundary condition (4.15) we get

$$u_n(0) + bv_n(0) = A_n, \quad n = 0, 1, \ldots, \tag{4.27}$$

where

$$A_n = \frac{1}{2\pi}\int_0^{2\pi} f(t)e^{-int}dt. \tag{4.28}$$

Substituting $v_n(x)$ and $u_n(x)$ from (4.23), (4.24) and (4.25) into (4.27), we get

$$c_0 = A_0, \tag{4.29}$$

$$c_n = \frac{i\mu\mu_0 n A_n}{\lambda_n + nb\mu\mu_0 i}, \quad n = 1, 2, \ldots. \tag{4.30}$$

Since $Re\lambda_n \neq 0$ for $n = 1, 2, \ldots$, the denominator of the fraction (4.30) is not equal to zero.

Substituting the constants $c_0, c_1, \ldots$ from (4.29) and (4.30) into (4.23) – (4.25) we get

$$u_0(x) = A_0, \quad v_0(x) = 0, \quad x > 0, \tag{4.31}$$

$$u_n(x) = \frac{\lambda_n A_n}{\lambda_n + inb\mu\mu_0}e^{\lambda_n x}, \quad v_n(x) = \frac{in\mu\mu_0 A_n}{\lambda_n + inb\mu\mu_0}e^{\lambda_n x}, \ x > 0. \tag{4.32}$$

From (4.16), (4.17), (4.31) and (4.32) we have

$$H_y(x,t) = A_0 + 2Re\sum_{n=1}^{\infty}\frac{\lambda_n A_n}{\lambda_n + inb\mu\mu_0}\exp(int + \lambda_n x), \tag{4.33}$$

$$E_z(x,t) = 2Re\sum_{n=1}^{\infty}\frac{in\mu\mu_0 A_n}{\lambda_n + inb\mu\mu_0}\exp(int + \lambda_n x). \tag{4.34}$$

Since the function $f(t)$ is infinitely differentiable and 2π-periodic, then for any positive integer m there exists a constant c (depending only on m and $f(t)$) such that

$$| A_n | \leq \frac{c}{n^m}, \quad n = 1, 2, \ldots. \tag{4.35}$$

It follows from (4.22) and (4.35) that the series (4.33) and (4.34) converge uniformly and admit term by term differentiation with respect to the variables

x and t. Thus, if a solution of the problem (4.10), (4.15) exists it will be defined by (4.33), (4.34). Substituting $H_y(x,t)$ and $E_z(x,t)$ from (4.33) and (4.34) into (4.10) and (4.15) we conclude that they satisfy the considered problem.

Hence, the problem (4.10), (4.15) is uniquely solvable and the solution is defined by (4.33) and (4.34).

Theorem 4.1 is proved.

Proof of Theorem 4.2. Let the boundary condition for system (4.10) have the form

$$E_z(0,t) = f(t). \tag{4.36}$$

Further, let $H_y(x,t)$ and $E_z(x,t)$ be the solutions of the problem (4.10), (4.36) belonging to the mentioned class and let (4.16), (4.17) be the Fourier series expansions of these functions with respect to the variable t. Then from (4.18) and (4.25) we have

$$v_n(0) = \frac{1}{2\pi}\int_0^{2\pi} E_z(0,t)e^{-int}dt, \quad n = 1,2,\ldots, \tag{4.37}$$

and

$$\int_0^{2\pi} E_z(0,t)dt = 0. \tag{4.38}$$

Substituting $E_z(0,t)$ from (4.36) into (4.37) and (4.38) we get

$$v_n(0) = A_n, \quad n = 1,2,\ldots, \tag{4.39}$$

and

$$\int_0^{2\pi} f(t)dt = 0, \tag{4.40}$$

where A_n is defined by (4.28).

Thus, the condition (4.40) is necessary for the solvability of the problem (4.10), (4.36).

Substituting $v_n(x)$ from (4.23) into (4.39) we get

$$c_n = A_n, \quad n = 1,2,\ldots. \tag{4.41}$$

From (4.16), (4.17), (4.23) – (4.25) and (4.41) we obtain

$$H_y(x,t) = c_0 + 2Re\sum_{n=1}^{\infty}\frac{A_n\lambda_n}{in\mu\mu_0}\exp(\lambda_n x + int), \tag{4.42}$$

$$E_z(x,t) = 2Re\sum_{n=1}^{\infty} A_n\exp(\lambda_n x + int). \tag{4.43}$$

Theorem 4.2 is proved.

4.3 A General Boundary Value Problem for the System (4.10) in Homogeneous Conducting Medium

Let R^2_+ be a conducting homogeneous medium. Consider the boundary value problem

$$k_1 \frac{\partial H_y(0,t)}{\partial x} + k_2 \frac{\partial E_z(0,t)}{\partial x} + k_3 H_y(0,t) + k_4 E_z(0,t) = f(t), \tag{4.44}$$

where k_1, k_2, k_3 and k_4 are given real constants, $k_1^2 + k_2^2 \neq 0$ and $f(t)$ is an infinitely differentiable 2π-periodic function.

The main goal of this paragraph is to obtain necessary and sufficient conditions in terms of the constants k_j $(j = 1, \ldots, 4)$ implying the unique solvability of the problem (4.10), (4.44).

One has the following:

Theorem 4.3 *For the unique solvability of the problem (4.10), (4.44) it is necessary and sufficient that*

$$k_3 \neq 0, \tag{4.45}$$

and

$$k_1(\sigma + in\varepsilon\varepsilon_0) + k_2\lambda_n - ik_3 \frac{\lambda_n}{n\mu\mu_0} + k_4 \neq 0, \quad n = 1, 2, \ldots. \tag{4.46}$$

Proof. Necessity. Let $k_3 = 0$. It is easy to see that $H_y(x,t) = const$ and $E_z(x,t) \equiv 0$ satisfy the homogeneous problem (4.10), (4.44). Hence the condition $k_3 \neq 0$ is necessary for the unique solvability of the problem (4.10), (4.44).

Now for any positive integer n let the condition

$$k_1(\sigma + in\varepsilon\varepsilon_0) + k_2\lambda_n - ik_3 \frac{\lambda_n}{n\mu\mu_0} + k_4 = 0 \tag{4.47}$$

hold. Then it is easy to check that the functions

$$H_y(x,t) = Re[\frac{c\lambda_n}{in\mu\mu_0} \exp(\lambda_n x + int)],$$
$$E_z(x,t) = Re[c \exp(\lambda_n x + int)],$$

where c is an arbitrary complex constant, satisfy the homogeneous problem (4.10), (4.44) (at $f(t) \equiv 0$). Hence the condition (4.46) is also necessary for the unique solvability of this problem.

Sufficiency. Let the conditions (4.45), (4.46) be fulfilled. We prove that the problem (4.10), (4.44) is uniquely solvable.

From (4.18) and (4.44) we have

$$k_1 u_n'(0) + k_2 v_n'(0) + k_3 u_n(0) + k_4 v_n(0) = A_n, \quad n = 0, 1, \ldots, \tag{4.48}$$

where A_n is defined by (4.28).

Substituting $u_n(x)$ and $v_n(x)$ from (4.23), (4.24) and (4.25) into (4.48), for $n = 1, 2, \ldots$ we get

$$c_0 k_3 = A_0, \quad c_n(k_1 \frac{\lambda_n^2}{in\mu\mu_0} + k_2\lambda_n + k_3 \frac{\lambda_n}{in\mu\mu_0} + k_4) = A_n. \tag{4.49}$$

Resolving equation (4.49) with respect to the constants c_0, $c_1, \ldots$ and taking into account (4.45), (4.46) and the equality $\lambda_n^2 = n\mu\mu_0(-n\varepsilon\varepsilon_0 + i\sigma)$, we get

$$c_0 = \frac{A_0}{k_3}, \tag{4.50}$$

$$c_n = \frac{A_n}{Q_n}, \quad n = 1, 2, \ldots, \tag{4.51}$$

where

$$Q_n = k_1(\sigma + in\varepsilon\varepsilon_0) + k_2\lambda_n - ik_3 \frac{\lambda_n}{n\mu\mu_0} + k_4, \quad n = 1, 2, \ldots. \tag{4.52}$$

By Taylor formula we have

$$\sqrt{1+x} = 1 + \frac{1}{2}x + \frac{1}{2!}\cdot\frac{1}{2}(\frac{1}{2} - 1)x^2 + \frac{1}{3!}\cdot\frac{1}{2}(\frac{1}{2} - 1)(\frac{1}{2} - 2)x^3 + x^3\gamma(x), \tag{4.53}$$

where $\gamma(x) \to 0$ as $x \to 0$.

From (4.22) and (4.53) we obtain

$$\lambda_n = -in\sqrt{\alpha} - \frac{1}{2}\beta\sqrt{\alpha} - \frac{\beta^2\sqrt{\alpha}i}{8n} - \frac{\beta^3\sqrt{\alpha}}{16n^2} + \frac{\gamma_n}{n^2}, \quad n = 1, 2, \ldots, \tag{4.54}$$

where

$$\gamma_n \to 0 \quad as \quad n \to +\infty, \quad \alpha = \varepsilon\varepsilon_0\mu\mu_0, \quad \beta = \frac{\sigma}{\varepsilon\varepsilon_0}.$$

Substituting λ_n from (4.54) into (4.52), we get

$$Q_n = a_1 n + a_0 + a_{-1}n^{-1} + a_{-2}n^{-2} + \frac{\delta_n}{n^2}, \tag{4.55}$$

where

$$a_1 = ik_1\varepsilon\varepsilon_0 - ik_2\sqrt{\alpha}, \tag{4.56}$$

$$a_0 = k_1\sigma - \frac{k_2\beta\sqrt{\alpha}}{2} - \frac{k_3\sqrt{\alpha}}{\mu\mu_0} + k_4, \tag{4.57}$$

$$a_{-1} = -\frac{k_2\beta^2\sqrt{\alpha}i}{8} + \frac{k_3\beta\sqrt{\alpha}i}{2\mu\mu_0}, \tag{4.58}$$

$$a_{-2} = -\frac{k_2\beta^3\sqrt{\alpha}}{16} - \frac{k_3\beta^2\sqrt{\alpha}}{8\mu\mu_0}, \tag{4.59}$$

and $\delta_n \to 0$ as $n \to +\infty$.

It follows from (4.55) – (4.59) that the numbers a_1, a_0, a_{-1}, and a_{-2} are equal to zero simultaneously if and only if $k_j = 0$ $(j = 1, 2, 3, 4)$. Since $k_1^2 + k_2^2 \neq 0$, at least one of the numbers a_1, a_0, a_{-1}, a_{-2} is not equal to zero. So from (4.46) and (4.55) we have

$$| Q_n | \geq \frac{c}{n^2}, \tag{4.60}$$

where c is a positive constant, independent of n.

Hence if the solution $H_y(x,t)$ and $E_z(x,t)$ exists, they will be defined by (4.16), (4.17) where the functions $u_n(x)$, $v_n(x)$ and the constants c_k $(k = 0, 1, \ldots)$ are defined by (4.23), (4.24), (4.25), (4.50) and (4.51). Using the estimates (4.35) and (4.60) one can check that the above defined functions $H_y(x,t)$ and $E_z(x,t)$ satisfy the problem (4.10), (4.59).

Theorem 4.3 is proved.

Now let the number k_3 in the condition (4.44) be equal to zero, i.e.

$$k_1\frac{\partial H_y(0,t)}{\partial x} + k_2\frac{\partial E_z(0,t)}{\partial x} + k_4E_z(0,t) = f(t). \tag{4.61}$$

It is easy to check that for any real constant c_0, $H_y(x,y) = c_0$ and $E_z(x,y) = 0$ satisfy the homogeneous problem (4.10), (4.61) (at $f(t) \equiv 0$).

In order to determine c_0 uniquely, we assume that $H_y(x,t)$ satisfies the additional condition

$$\frac{1}{2\pi}\int_0^{2\pi} H_y(0,t)dt = \delta_0, \tag{4.62}$$

where δ_0 is a given real number.

From (4.18) (for $n = 0$) and (4.61) we have

$$k_1u_0'(0) + k_2v_0'(0) + k_4v_0(0) = \frac{1}{2\pi}\int_0^{2\pi} f(t)dt. \tag{4.63}$$

Substituting $u_0(x)$ and $v_0(x)$ from (4.25) into (4.63), we get

$$\int_0^{2\pi} f(t)dt = 0. \qquad (4.64)$$

Hence condition (4.64) is necessary for the solvability of the problem (4.10), (4.61).

One has the following

Theorem 4.4 *Let the necessary condition (4.64) be satisfied. Then the problem (4.10), (4.61) with the additional condition (4.62) is uniquely solvable if and only if the condition (4.46) at $k_3 = 0$ is satisfied.*

Proof is similar to that of Theorem 4.3, we note only that the formula (4.50) should be replaced by

$$c_0 = \delta_0. \qquad (4.65)$$

Since in (4.55) some of the numbers a_1, a_0, a_{-1}, a_{-2} are not equal to zero, then for large values of n one has the inequality (4.60). This means that the numbers Q_n can be equal to zero for a finite number of indices $n = m_1, \ldots, m_j$, i.e.

$$Q_{m_k} = 0, \quad k = 1, \ldots, j, \qquad (4.66)$$
$$Q_n \neq 0, \quad n \in N \setminus \{m_1, \ldots, m_j\}, \qquad (4.67)$$

where Q_n is defined by (4.52) and N is the set of positive integers.

One has the following

Theorem 4.5 *If $k_3 \neq 0$ and the conditions (4.66) and (4.67) are satisfied, then the homogeneous problem (4.10), (4.44) has just $2j$ linear independent solutions and the corresponding non-homogeneous problem is solvable if and only if*

$$\int_0^{2\pi} f(t)\cos m_k t dt = 0, \quad \int_0^{2\pi} f(t)\sin m_k t dt = 0, \quad k = 1, \ldots, j. \qquad (4.68)$$

Proof. From (4.18) and (4.44) we have

$$k_1 u_n'(0) + k_2 v_n'(0) + k_3 u_n(0) + k_4 v_n(0) = A_n, \quad n = 0, 1, \ldots, \qquad (4.69)$$

where A_n is defined by (4.28).

Substituting $u_n(x)$ and $v_n(x)$ from (4.23), (4.24) and (4.25) into (4.69), we get

$$k_3 c_0 = A_0, \tag{4.70}$$
$$Q_n c_n = A_n, \quad n = 1, 2, \ldots, \tag{4.71}$$

where Q_n is defined by (4.52).

Using the conditions (4.66), (4.67) from (4.70), (4.71) we get

$$c_0 = \frac{A_0}{k_3}, \tag{4.72}$$
$$c_n = \frac{A_n}{Q_n}, \quad n \in N, \quad n \neq m_1, \ldots, m_j, \tag{4.73}$$
$$A_{m_k} = 0, \quad k = 1, \ldots, j, \tag{4.74}$$

and c_{m_k} $(k = 1, \ldots, j)$ are arbitrary complex constants.

The conditions (4.74) coincide with (4.68). Hence the conditions (4.68) are necessary for the solvability of the problem (4.10), (4.44). Let (4.68) be satisfied. Then, if the solution exists, it will be represented as in (4.16), (4.17) where u_n and v_n are defined by (4.23) – (4.25) and c_n for $n \neq m_1, \ldots, m_j$ are defined by (4.73). In (4.23) and (4.24) the remaining constants $c_{m_1}, \ldots, c_{m_j}$ were arbitrarily chosen. It is easy to show that the functions $H_y(x,t)$ and $E_z(x,t)$ constructed in this way satisfy the considered problem.

The proof of Theorem 4.5 immediately follows from the obtained formula for solution. The linearly independent solutions of the homogeneous problem are the following functions

$$H_y(x,t) = Re\left[\frac{\lambda_n}{in\mu\mu_0}\exp(\lambda_n x + nit)\right],$$
$$E_z(x,t) = Re\exp(\lambda_n x + nit)$$

and

$$H_y(x,t) = Re\left[\frac{\lambda_n}{n\mu\mu_0}\exp(\lambda_n x + nit)\right],$$
$$E_z(x,t) = Re[i\exp(\lambda_n x + nit)],$$

where $n = m_1, \ldots, m_j$.

Let $k_3 = 0$ and the conditions (4.66), (4.67) be satisfied. Then one has the following

Theorem 4.6 *The homogeneous problem (4.10), (4.48) has just $2j+1$ linearly independent solutions and the non-homogeneous problem is solvable if and only if the conditions (4.64) and (4.68) are satisfied.*

Proof is similar to that of Theorem 4.5. In this case, to the above mentioned linearly independent solutions of the homogeneous problem we add one more solution $H_y(x,t) = 1$, $E_z(x,t) = 0$.

Now using Theorem 4.3 and the conditions (4.46), we are going to indicate some simple sufficient conditions in terms of the coefficients k_1, k_2, k_3, k_4 providing unique solvability of the problem (4.10), (4.44).

According to Theorem 4.3, for unique solvability of this problem it is necessary that $k_3 \neq 0$. So, without loss of generality, we can assume that $k_3 > 0$.

Now we indicate sufficient conditions in terms of k_1, k_2, k_3, k_4 yielding the condition (4.46) for any $k_3 > 0$. The condition (4.46) can be written as

$$Q_n \neq 0, \quad n = 1, 2, \ldots, \tag{4.75}$$

or

$$\frac{Q_n}{\lambda_n} \neq 0, \quad n = 1, 2, \ldots, \tag{4.76}$$

where Q_n is defined by (4.52). Since

$$\lambda_n^2 = n\mu\mu_0(-n\varepsilon\varepsilon_0 + i\sigma), \tag{4.77}$$

we have

$$\frac{Q_n}{\lambda_n} = -\frac{ik_1\lambda_n}{n\mu\mu_0} + k_2 - \frac{ik_3}{n\mu\mu_0} + \frac{k_4}{\lambda_n}. \tag{4.78}$$

According to (4.22), $\lambda_n = -a_n - ib_n$, where a_n and b_n are positive constants depending on n.

From (4.22), (4.52) and (4.78) we have

$$ImQ_n = k_1 n\varepsilon\varepsilon_0 - k_2 b_n + \frac{k_3 a_n}{n\mu\mu_0}, \tag{4.79}$$

$$Im\frac{Q_n}{\lambda_n} = \frac{k_1 a_n}{n\mu\mu_0} - \frac{k_3}{n\mu\mu_0} + \frac{b_n k_4}{a_n^2 + b_n^2}. \tag{4.80}$$

From the relations (4.79), (4.80) we have

$$ImQ_n > 0 \quad for \quad k_1 \geq 0, \quad k_2 \leq 0, \quad k_3 > 0, \tag{4.81}$$

$$Im\frac{Q_n}{\lambda_n} < 0 \quad for \quad k_1 \leq 0, \quad k_3 > 0, \quad k_4 \leq 0. \tag{4.82}$$

Thus, if $k_1 \geq 0$, $k_2 \leq 0$, $k_3 > 0$ or $k_1 \leq 0$, $k_3 > 0$, $k_4 \leq 0$, the conditions (4.45), (4.46) of Theorem 4.3 are satisfied. So we get the following

Corollary 4.1 *Let either* $k_3 > 0$, $k_1 \geq 0$, $k_2 \leq 0$ *or* $k_3 > 0$, $k_1 \leq 0$, $k_4 \leq 0$, *then the problem (4.10), (4.44) is uniquely solvable.*

Now let either $k_3 = 0,\ k_1k_2 \leq 0$ or $k_3 = 0,\ k_1k_4 \geq 0$. Then from (4.79) and (4.80) it follows that the conditions (4.75) are satisfied. Hence, from Theorem 4.4 we get the following

Corollary 4.2 *Let either $k_1k_2 \leq 0$ or $k_1k_4 \geq 0$, then the homogeneous problem (4.10), (4.61) has only one solution $H_y(x,t) = const,\ E_z(x,t) = 0$ and the corresponding non-homogeneous problem is solvable if and only if the condition (4.64) is satisfied.*

Now we consider some special cases of the general boundary value problem (4.10), (4.44), where the left-hand side of the boundary condition (4.44) consists of two terms, i.e. instead of (4.44) we have one of the following conditions:

$$\frac{\partial H_y(0,t)}{\partial x} + k_3 H_y(0,t) = f(t), \tag{4.83}$$

$$\frac{\partial E_z(0,t)}{\partial x} + k_3 H_y(0,t) = f(t), \tag{4.84}$$

$$k_1\frac{\partial H_y(0,t)}{\partial x} + k_2\frac{\partial E_z(0,t)}{\partial x} = f(t), \tag{4.85}$$

$$\frac{\partial H_y(0,t)}{\partial x} + k_4 E_z(0,t) = f(t), \tag{4.86}$$

$$\frac{\partial E_z(0,t)}{\partial x} + k_4 E_z(0,t) = f(t), \tag{4.87}$$

where $k_j\ (j = 1,\ldots,4)$ are real constants satisfying

$$k_1^2 + k_2^2 \neq 0, \quad k_3 \neq 0, \quad k_4 \neq 0. \tag{4.88}$$

Now we verify (4.46) for these boundary conditions.
The condition (4.46) for the problem (4.10), (4.83) takes the form

$$(\sigma + in\varepsilon\varepsilon_0) - ik_3\frac{\lambda_n}{n\mu\mu_0} \neq 0, \quad n = 1,2,\ldots. \tag{4.89}$$

Since $\lambda_n = -a_n - ib_n,\ a_n > 0,\ b_n > 0$ and

$$\lambda_n^2 = n\mu\mu_0(-n\varepsilon\varepsilon_0 + i\sigma), \tag{4.90}$$

the condition (4.89) can be written as

$$\lambda_n + k_3 \neq 0, \quad n = 1,2,\ldots. \tag{4.91}$$

By virtue of

$$Im(\lambda_n + k_3) = -b_n \neq 0, \quad n = 1,2,\ldots \tag{4.92}$$

the condition (4.91) is always satisfied.

Thus, from Theorem 4.3 we obtain the following

Corollary 4.3 *The problem (4.10), (4.83) is uniquely solvable.*

The condition (4.46) for the problem (4.10), (4.84) takes the form

$$\lambda_n - ik_3 \frac{\lambda_n}{n\mu\mu_0} \neq 0, \quad n = 1, 2, \ldots. \tag{4.93}$$

Since $\lambda_n \neq 0$, these inequalities hold for any positive integer n. Hence from Theorem 4.3 we obtain

Corollary 4.4 *The problem (4.10), (4.84) is uniquely solvable.*

In a similar way one can prove the following statement

Corollary 4.5 *The homogeneous problem (4.10), (4.85) has only one solution $H_y(x,t) = const$, $E_z(x,t) = 0$ and the non-homogeneous problem (4.2), (4.85) is solvable if and only if the condition (4.64) is satisfied.*

The assertion of Corollary 4.5 remains valid for the problems (4.10), (4.86) and (4.10), (4.87).

4.4 Boundary Value Problems for the System (4.10) in $x > 0$ Consisting of Two Homogeneous Strata

Let $x > 0$ consist of two homogeneous conducting strata $0 \leq x \leq x_1$ and $x_1 < x < \infty$. Then the quantities ε, μ and σ in equation (4.10) are piecewise constant. We have

$$\varepsilon = \varepsilon_1, \quad \mu = \mu_1, \quad \sigma = \sigma_1, \quad \text{for} \quad 0 \leq x < x_1, \tag{4.94}$$

$$\varepsilon = \varepsilon_2, \quad \mu = \mu_2, \quad \sigma = \sigma_2, \quad \text{for} \quad x > x_1, \tag{4.95}$$

where ε_1, ε_2, μ_1, μ_2, σ_1, and σ_2 are some positive constants.

Based on the physical considerations we assume that
1) the solutions $H_y(x,t)$ and $E_z(x,t)$ satisfy (4.10) in the domains $0 < x < x_1$, $-\infty < t < +\infty$ and $x > x_1$, $-\infty < t < +\infty$;
2) $H_y(x,t)$ and $E_z(x,t)$ are bounded and continuous in the domain $\overline{R^2_+} = \{(x,t);\ x \geq 0,\ -\infty < t < +\infty\}$ and are 2π-periodic on t;
3) H_y and E_z are infinitely differentiable in the domains $0 < x < x_1$, $t \in (-\infty, +\infty)$ and $x > x_1$, $t \in (-\infty, +\infty)$.

We take the boundary condition in the form

$$H_y(0,t) = kE_z(0,t) + f(t), \tag{4.96}$$

where k is a non-negative number and $f(t)$ satisfies the same conditions as in (4.13).

One has the following:

Theorem 4.7 *The problem (4.10), (4.96) in two strata medium is uniquely solvable.*

Proof. Let $H_y(x,t)$ and $E_z(x,t)$ be a solution of the system (4.10) and $u_n(x)$ and $v_n(x)$ be the Fourier coefficients of these functions, defined by (4.18). Then for the system of equations (4.19) in two strata medium we get

$$\frac{du_n(x)}{dx} = (in\varepsilon_0\varepsilon_1 + \sigma_1)v_n(x), \ \frac{dv_n(x)}{dx} = in\mu_0\mu_1 u_n(x), \ 0 < x < x_1, \quad (4.97)$$

$$\frac{du_n(x)}{dx} = (in\varepsilon_0\varepsilon_2 + \sigma_2)v_n(x), \ \frac{dv_n(x)}{dx} = in\mu_0\mu_2 u_n(x), \quad x > x_1. \quad (4.98)$$

Let $n \geq 1$. For $0 < x < x_1$ the general solution of (4.97) is defined by

$$v_n(x) = c_{n_1}e^{\lambda_n x} + c_{n_2}e^{-\lambda_n x}, \quad u_n(x) = \frac{\lambda_n}{in\mu_1\mu_0}(c_{n_1}e^{\lambda_n x} - c_{n_2}e^{-\lambda_n x}), \quad (4.99)$$

where c_{n_1} and c_{n_2} are arbitrary constants and

$$\lambda_n = -\sqrt{n\alpha_1(-n + i\beta_1)} = -a_{n_1} - ib_{n_1}, \quad (4.100)$$

with

$$a_{n_1} = \frac{\beta_1\sqrt{\alpha_1}}{\sqrt{2}\sqrt{\sqrt{1+\beta_1^2 n^{-2}} + 1}}, \quad b_{n_1} = \frac{n\sqrt{\alpha_1}}{\sqrt{2}}\sqrt{1 + \sqrt{1+\beta_1^2 n^{-2}}} \quad (4.101)$$

and

$$\alpha_1 = \varepsilon_0\mu_0\varepsilon_1\mu_1, \quad \beta_1 = \frac{\sigma_1}{\varepsilon_1\varepsilon_0}.$$

Since the interval $(0, x_1)$ is bounded, all the solutions of the system (4.97) on this interval are uniformly bounded. The general solution of the system (4.98), bounded in the domain $x > x_1$ is defined by

$$v_n(x) = c_{n_3}e^{\nu_n x}, \quad u_n(x) = \frac{\nu_n}{in\mu_2\mu_0}c_{n_3}e^{\nu_n x}, \quad x > x_1, \ n \geq 1, \quad (4.102)$$

where c_{n_3} is an arbitrary constant and

$$\nu_n = -\sqrt{n\alpha_2(-n + i\beta_2)} = -a_{n_2} - ib_{n_2}, \quad (4.103)$$

with

$$a_{n_2} = \frac{\beta_2\sqrt{\alpha_2}}{\sqrt{2}\sqrt{\sqrt{1+\beta_2^2 n^{-2}}+1}}, \quad b_{n_2} = \frac{n\sqrt{\alpha_2}}{\sqrt{2}}\sqrt{1+\sqrt{1+\beta_2^2 n^{-2}}} \tag{4.104}$$

and

$$\alpha_2 = \varepsilon_0\mu_0\varepsilon_2\mu_2, \quad \beta_2 = \frac{\sigma_2}{\varepsilon_0\varepsilon_2}.$$

Since $H_y(x,t)$ and $E_z(x,t)$ are continuous at the boundary $x = x_1$ of two strata, the functions $u_n(x)$ and $v_n(x)$ are continuous at $x = x_1$ as well.

Now let $n = 0$. Then resolving the system (4.97) and (4.98) in the class of bounded functions we get

$$v_0(x) = c_{01}, \quad u_0(x) = c_{01}\sigma_1 x + c_{02}, \quad 0 < x < x_1, \tag{4.105}$$

$$v_0(x) = 0, \quad u_0(x) = c_{03}, \quad x > x_1, \tag{4.106}$$

where c_{01}, c_{02} and c_{03} are arbitrary real constants.

Since $v_0(x)$ and $u_0(x)$ are continuous at the point $x = x_1$, from (4.105) and (4.106) we get

$$v_0(x) = 0, \quad u_0(x) = c_{02}, \quad x > 0. \tag{4.107}$$

From (4.18) and (4.96) we have

$$u_n(0) = kv_n(0) + A_n, \quad n = 0,1,\ldots, \tag{4.108}$$

where the constants A_n are defined by (4.28).

Substituting $u_n(x)$ and $v_n(x)$ from (4.99) into (4.108) we get

$$\frac{\lambda_n}{in\mu_1\mu_0}(c_{n_1} - c_{n_2}) = k(c_{n_1} + c_{n_2}) + A_n, \quad n = 1,2,\ldots. \tag{4.109}$$

From (4.99) and (4.102) and from continuity of $u_n(x)$ and $v_n(x)$ at x_1 we get

$$c_{n_1}e^{\lambda_n x_1} + c_{n_2}e^{-\lambda_n x_1} = c_{n_3}e^{\nu_n x_1}, \quad n = 1,2,\ldots, \tag{4.110}$$

$$\mu_2\lambda_n(c_{n_1}e^{\lambda_n x_1} - c_{n_2}e^{-\lambda_n x_1}) = \mu_1\nu_n c_{n_3}e^{\nu_n x_1}, \quad n = 1,2,\ldots. \tag{4.111}$$

Resolving the system of equations (4.109) – (4.111) with respect to the constants c_{n1}, c_{n2} and c_{n3} for $n \geq 1$ we get

$$c_{n3} = \frac{2i\lambda_n\mu_1\mu_2\mu_0 A_n}{nQ_n}\exp(-\nu_n x_1), \tag{4.112}$$

$$c_{n1} = \frac{(\mu_2\lambda_n + \nu_n\mu_1)\mu_0\mu_1 i}{nQ_n}A_n\exp(-\lambda_n x_1), \tag{4.113}$$

$$c_{n2} = \frac{(\mu_2\lambda_n - \nu_n\mu_1)\mu_0\mu_1 i}{nQ_n}A_n\exp(\lambda_n x_1), \tag{4.114}$$

where

$$\begin{aligned} Q_n &= n^{-2}(\lambda_n - kn\mu_1\mu_0 i)(\mu_2\lambda_n + \mu_1\nu_n)\exp(-\lambda_n x_1) \\ &- n^{-2}(\lambda_n + kn\mu_1\mu_0 i)(\lambda_n\mu_2 - \mu_1\nu_n)\exp(\lambda_n x_1). \end{aligned} \tag{4.115}$$

From (4.115) we have

$$\begin{aligned} n^2 \mid Q_n \mid &\geq \mid \lambda_n - kn\mu_1\mu_0 i \mid\mid \mu_2\lambda_n + \nu_n\mu_1 \mid \exp(a_{n1}x_1) \\ &- \mid \lambda_n + kn\mu_1\mu_0 i \mid\mid \mu_2\lambda_n - \nu_n\mu_1 \mid \exp(-a_{n1}x_1) \mid . \end{aligned} \tag{4.116}$$

Since $x_1 > 0$, $k \geq 0$, $Re\lambda_n < 0$, $Re\nu_n < 0$, $Im\lambda_n < 0$, $Im\nu_n < 0$, and $a_{n1} > 0$, for $n \geq 1$ we get

$$\mid \lambda_n - ik\mu_1\mu_0 n \mid > \mid \lambda_n + ik\mu_1\mu_0 n \mid, \tag{4.117}$$

$$\mid \mu_2\lambda_n + \mu_1\nu_n \mid > \mid \mu_2\lambda_n - \mu_1\nu_n \mid, \tag{4.118}$$

$$\exp(a_n x_1) > \exp(-a_n x_1). \tag{4.119}$$

The inequalities (4.117) – (4.119) imply

$$Q_n \neq 0, \quad n = 1, 2, \ldots . \tag{4.120}$$

Further, in view of (4.100), (4.101), (4.103) and (4.104) we can write

$$\lim_{n\to\infty} \frac{\lambda_n}{n} = -i\sqrt{\alpha_1}, \quad \lim_{n\to+\infty} a_{n1} = \frac{\beta_1\sqrt{\alpha_1}}{\sqrt{2}}, \tag{4.121}$$

$$\lim_{n\to\infty} \frac{\nu_n}{n} = -i\sqrt{\alpha_2}. \tag{4.122}$$

Using the limits (4.121) and (4.122) we get

$$\begin{aligned} &\lim_{n\to+\infty} [n^{-2} \mid \lambda_n - \mu_1\mu_0 kin \mid \mid \mu_2\lambda_n + \nu_n\mu_1 \mid \exp(a_{n1}x_1) \\ &\quad -n^{-2} \mid \lambda_n + \mu_1\mu_0 kni \mid\mid \mu_2\lambda_n - \mu_1\nu_n \mid \exp(-a_{n1}x_1)] \\ &\quad = \mid k\mu_1\mu_0 + \sqrt{\alpha_1} \mid\mid \mu_2\sqrt{\alpha_1} + \mu_1\sqrt{\alpha_2} \mid \exp(\frac{\beta_1\sqrt{\alpha_1}x_1}{\sqrt{2}}) \\ &- \mid k\mu_1\mu_0 - \sqrt{\alpha_1} \mid\mid \mu_2\sqrt{\alpha_1} - \mu_1\sqrt{\alpha_2} \mid \exp(-\frac{\beta_1\sqrt{\alpha_1}x_1}{\sqrt{2}}). \end{aligned} \tag{4.123}$$

For $k \geq 0$ the right-hand side of (4.123) is positive. Hence from (4.116), (4.120) and (4.123) we have

$$\mid Q_n \mid \geq \tilde{Q}, \quad n = 1, 2, \ldots, \tag{4.124}$$

where $\tilde{Q}$ is some positive constant which is independent of n.

From (4.112) – (4.116) and (4.121), (4.122) and (4.124) we obtain

$$| c_{nj} | \leq c_0 | A_n |, \quad j = 1,2,3; \quad n = 1,2,\ldots, \tag{4.125}$$

where c_0 is some positive constant, independent of n.

Substituting the functions $u_0(x)$ and $v_0(x)$ from (4.107) into (4.108) at $n = 0$ we get

$$c_{02} = A_0. \tag{4.126}$$

Hence

$$u_0(x) = A_0, \quad v_0(x) = 0, \quad 0 < x < \infty. \tag{4.127}$$

Therefore, if a solution of the problem (4.10), (4.96) in two strata medium $x > 0$ exists it will be defined by (4.16), (4.17) where the functions $u_n(x)$ and $v_n(x)$ and the constants c_{n1}, c_{n2} and c_{n3} ($n = 1,2,\ldots$) are defined by (4.99), (4.102), (4.112) – (4.114) and $u_0(x)$ and $v_0(x)$ are defined by (4.127).

On the other hand, using the estimates (4.124) and (4.125) it is easy to check that the functions $H_y(x,t)$ and $E_z(x,t)$ defined by (4.16) and (4.17) satisfy the problem (4.10), (4.96).

Theorem 4.7 is proved.

Consider now the boundary condition

$$E_z(0,t) = f(t), \tag{4.128}$$

where $f(t)$ is the same function as in (4.13).

From (4.18) and (4.128) we have

$$v_n(0) = A_n, \quad n = 0,1,\ldots. \tag{4.129}$$

From (4.107) and (4.129) we get

$$A_0 = \frac{1}{2\pi} \int_0^{2\pi} f(t)dt = 0. \tag{4.130}$$

Thus, the condition (4.130) is necessary for the solvability of the problem (4.10), (4.128) in two-strata conducting medium $x > 0$. On the other hand, it is clear that $H_y(x,t) = c_0$, $E_z(x,t) = 0$ is a solution of the homogeneous problem (4.10), (4.128) (c_0 is an arbitrary real constant). This means that the problem (4.10), (4.128) is not uniquely solvable. We impose an additional condition of the form

$$\frac{1}{2\pi} \int_0^{2\pi} H_y(0,t)dt = \gamma_0, \tag{4.131}$$

where γ_0 is a given real constant.

One has the following theorem.

Theorem 4.8 *Let condition (4.130) be satisfied. Then the problem (4.10), (4.128) with additional condition (4.131) is uniquely solvable.*

Proof is similar to that of Theorem 4.7.

Now we write a solution of the problem (4.10), (4.128) in the explicit form. Let

$$u_0(x) = \gamma_0, \quad v_0(x) = 0, \quad x \geq 0, \tag{4.132}$$

$$l_n = (\mu_2\lambda_n + \mu_1\nu_n)\exp(-\lambda_n x_1) + (\mu_2\lambda_n - \mu_1\nu_n)\exp(\lambda_n x_1), \tag{4.133}$$

$$c_{n1} = \frac{\mu_2\lambda_n + \mu_1\nu_n}{l_n} A_n \exp(-\lambda_n x_1), \tag{4.134}$$

$$c_{n2} = \frac{\mu_2\lambda_n - \mu_1\nu_n}{l_n} A_n \exp(\lambda_n x_1), \tag{4.135}$$

$$c_{n3} = \frac{2\mu_2 A_n \lambda_n}{l_n} \exp(-\nu_n x_1). \tag{4.136}$$

It is easy to check that

$$l_n \neq 0, \quad n = 1, 2, \ldots. \tag{4.137}$$

A solution of the problem (4.10), (4.128) with additional condition (4.131) in two strata conducting medium $x > 0$ is defined by (4.16), (4.17), where $u_0(x) = \gamma_0$, $v_0(x) = 0$ and $u_n(x)$, $v_n(x)$, c_{n1}, c_{n2}, c_{n3} $(n = 1, 2, \ldots)$ are defined by (4.99) – (4.102), (4.134) – (4.136).

We take now the boundary condition in the form

$$\frac{\partial H_y(0,t)}{\partial x} = kH_y(0,t) + f(t), \tag{4.138}$$

where k is a positive number and $f(t)$ satisfies the same conditions as in (4.13).

One has the following

Theorem 4.9 *The problem (4.10), (4.138) in two strata medium $x > 0$ is uniquely solvable.*

Proof. From (4.18) and (4.138) we have

$$u_n'(0) = ku_n(0) + A_n, \quad n = 0, 1, 2, \ldots, \tag{4.139}$$

where A_n is defined by (4.28).

Substituting $u_n(x)$ and $v_n(x)$ from (4.99) into (4.139) we get

$$\frac{\lambda_n^2}{in\mu_1\mu_0}(c_{n1} + c_{n2}) = \frac{k\lambda_n}{in\mu_1\mu_0}(c_{n1} - c_{n2}) + A_n. \tag{4.140}$$

Resolving the system (4.110), (4.111) and (4.140) with respect to c_{n1}, c_{n2} and c_{n3} we get

$$c_{n1} = \frac{A_n\mu_1\mu_0 in(\mu_2\lambda_n + \mu_1\nu_n)}{\omega_n\lambda_n}\exp(-\lambda_n x_1), \tag{4.141}$$

$$c_{n2} = \frac{A_n\mu_1\mu_0 in(\mu_2\lambda_n - \mu_1\nu_n)}{\omega_n\lambda_n}\exp(\lambda_n x_1), \tag{4.142}$$

$$c_{n3} = \frac{2n\mu_0\mu_1\mu_2 A_n i}{\omega_n}\exp(-\nu_n x_1), \quad n = 1, 2, \ldots, \tag{4.143}$$

where

$$\begin{aligned}\omega_n &= (\lambda_n - k)(\mu_2\lambda_n + \mu_1\omega_n)e^{-\lambda_n x_1}\\ &+ (\lambda_n + k)(\mu_2\lambda_n - \mu_1\omega_n)e^{\lambda_n x_1}, \quad n = 1, 2, \ldots.\end{aligned}$$

As for the inequality (4.124), we can show that

$$|\,\omega_n\,| \geq cn^2, \quad n = 1, 2, \ldots,$$

where c is some positive constant.

Substituting $u_0(x)$ and $v_0(x)$ from (4.107) into (4.139) at $n = 0$ we get

$$u_0(x) = -\frac{A_0}{k}, \quad v_0(x) = 0, \quad x \geq 0. \tag{4.144}$$

Hence if the problem (4.10), (4.138) is solvable in two strata medium $x > 0$, then the solution will be defined by (4.16), (4.17), where $u_0(x)$ and $v_0(x)$ are defined by (4.144) and the functions $u_n(x)$, $v_n(x)$ and the constants c_{n1}, c_{n2} and c_{n3} $n = 1, 2, \ldots$ are defined by (4.99), (4.102), (4.141) – (4.143). It is easy to check that the functions $H_y(x,t)$ and $E_z(x,t)$ satisfy the problem (4.10), (4.138) in two strata medium $x > 0$.

Theorem 4.9 is proved.

We take now the boundary condition in the form

$$\frac{\partial E_z(x,t)}{\partial x} = kH_y(0,t) + f(t), \tag{4.145}$$

where k is a nonzero real constant.

From (4.18) and (4.145) we have

$$v_n'(0) = ku_n(0) + A_n, \quad n = 0, 1, 2, \ldots. \tag{4.146}$$

Substituting $v_n(x)$ and $u_n(x)$ from (4.99) into (4.146) we get

$$\lambda_n(c_{n1} - c_{n2}) = \frac{\lambda_n k}{in\mu_1\mu_0}(c_{n1} - c_{n2}) + A_n \tag{4.147}$$

or

$$c_{n1} - c_{n2} = \frac{\mu_1 \mu_0 n A_n i}{\lambda_n (i n \mu_1 \mu_0 - k)}. \tag{4.148}$$

Denote

$$\tilde{l}_n = n^{-1}(\mu_1 \nu_n + \lambda_n \mu_2) \exp(-\lambda_n x_1) + n^{-1}(\mu_1 \nu_n - \lambda_n \mu_2) \exp(\lambda_n x_1). \tag{4.149}$$

Then

$$| \tilde{l}_n | \geq [| \mu_1 \nu_n + \lambda_n \mu_2 | - | \nu_n \mu_1 - \lambda_n \mu_2 |] n^{-1}. \tag{4.150}$$

From here, as in the inequality (4.124), we get

$$| \tilde{l}_n | \geq c, \quad n = 1, 2, \ldots, \tag{4.151}$$

where c is a real positive constant, independent of n.

Resolving the system of equations (4.110), (4.111) and (4.147) with respect to c_{n1}, c_{n2} and c_{n3} for $n = 1, 2, \ldots$ we get

$$c_{n3} = \frac{2i\mu_0 \mu_1 \mu_2 A_n}{\tilde{l}_n (i n \mu_1 \mu_0 - k)} \exp(-\nu_n x_1), \tag{4.152}$$

$$c_{n1} = \frac{i A_n \mu_1 \mu_0 (\mu_2 \lambda_n + \mu_1 \nu_n)}{\tilde{l}_n (i n \mu_1 \mu_0 - k) \lambda_n} \exp(-\lambda_n x_1), \tag{4.153}$$

$$c_{n2} = \frac{i A_n \mu_1 \mu_0 (\mu_2 \lambda_n - \mu_1 \nu_n)}{\tilde{l}_n (i n \mu_1 \mu_0 - k) \lambda_n} \exp(\lambda_n x_1). \tag{4.154}$$

As in (4.144), we have

$$u_0(x) = -\frac{A_0}{k}, \quad v_0(x) = 0, \quad x \geq 0. \tag{4.155}$$

Hence one has the following:

Theorem 4.10 *The problem (4.10), (4.145) in two strata medium $x > 0$ is uniquely solvable and the solution is defined by (4.16), (4.17), where $u_0(x)$ and $v_0(x)$ are defined by (4.155), while the functions $u_n(x)$ and $v_n(x)$ and the constants c_{n1}, c_{n2} and c_{n3} are defined by (4.99), (4.102), (4.152) – (4.154).*

Consider now the boundary condition

$$\frac{\partial E_z(0,t)}{\partial x} = k \frac{\partial H_y(0,t)}{\partial x} + f(t), \tag{4.156}$$

where k is a non-negative number.

From (4.11) and (4.156) we have

$$v'_n(0) = k u'_n(0) + A_n, \quad n = 0, 1, \ldots. \tag{4.157}$$

Substituting $u_n(x)$ and $v_n(x)$ from (4.99) into (4.157) for $n \geq 1$ we get

$$\lambda_n(c_{n1} - c_{n2}) = \frac{k\lambda_n^2}{in\mu_1\mu_0}(c_{n1} + c_{n2})A_n, \quad n = 1, 2, \ldots. \tag{4.158}$$

Resolving the system of equations (4.110), (4.111) and (4.158) with respect to c_{n1}, c_{n2} and c_{n3} we get

$$c_{n3} = \frac{2A_n i\mu_0\mu_1\mu_2}{n\omega_n}\exp(-\nu_n x_1), \tag{4.159}$$

$$c_{n1} = \frac{A_n\mu_1\mu_0 i}{n\omega_n}.\frac{\mu_2\lambda_n + \nu_n\mu_1}{\lambda_n}\exp(-\lambda_n x_1), \tag{4.160}$$

$$c_{n2} = \frac{A_n\mu_1\mu_0 i}{n\omega_n}.\frac{\mu_2\lambda_n - \nu_n\mu_1}{\lambda_n}\exp(\lambda_n x_1), \tag{4.161}$$

where

$$\begin{aligned} \omega_n &= n^{-2}(\mu_2\lambda_n + \nu_n\mu_1)(in\mu_1\mu_0 - k\lambda_n)e^{-\lambda_n x_1} \\ &- n^{-2}(in\mu_1\mu_0 + k\lambda_n)(\mu_2\lambda_n - \nu_n\mu_1)e^{\lambda_n x_1}. \end{aligned} \tag{4.162}$$

Substituting $u_0(x)$ and $v_0(x)$ from (4.107) into (4.157) at $n = 0$ we get

$$u_0(x) = c_0, \quad v_0(x) = 0, \quad x \geq 0, \tag{4.163}$$

$$A_0 = 0, \tag{4.164}$$

where c_0 is an arbitrary constant.

Thus, we have

Theorem 4.11 *The homogeneous problem (4.10), (4.156) in two strata conducting medium $x > 0$ has one solution $H_y(x,t) = const$, $E_z(x,t) = 0$ and the corresponding non-homogeneous problem is solvable if and only if the condition (4.164) is satisfied. A solution of this problem is defined by (4.16), (4.17), where $u_0(x)$ and $v_0(x)$ are defined by (4.163) and the functions $u_n(x)$, $v_n(x)$ and the constants $c_{n_1}, c_{n_2}, c_{n_3}$, $(n = 1, 2, \ldots)$ are defined by (4.99), (4.102), (4.159) - (4.161).*

Consider now a more general boundary condition

$$k_1\frac{\partial H_y(0,t)}{\partial x} - k_2\frac{\partial E_z(0,t)}{\partial x} + k_3 E_z(0,t) - k_4 H_y(0,t) = f(t), \tag{4.165}$$

where k_1, k_2, k_3 and k_4 are real constants and $k_1^2 + k_2^2 \neq 0$.

Using the above arguments we can prove the following two assertions.

Theorem 4.12 *If $k_j \geq 0$ $(j = 1, 2, 3)$ and $k_4 > 0$, then the problem (4.10), (4.165) in two strata medium $x > 0$ is uniquely solvable.*

Theorem 4.13 *If $k_j \geq 0$ $(j = 1,2,3)$ and $k_4 = 0$, then the homogeneous problem (4.10), (4.165) in two strata medium $x > 0$ has one solution $H_y(x,t) = const$, $E_z(x,t) = 0$ and the non-homogeneous problem is solvable if and only if*

$$\int_0^{2\pi} f(t)dt = 0. \tag{4.166}$$

4.5 On Fredholmity of General Boundary Value Problem for Equation (4.10) in Two Strata Medium $x > 0$

Definition 4.1. The problem (4.10), (4.165) is called Fredholmian if the following two conditions hold:
1) the homogeneous problem (4.10), (4.165) has at most a finite number of linearly independent solutions;
2) the non-homogeneous problem (4.10), (4.165) is solvable if and only if the function $f(t)$ satisfies a finite number of conditions of the form

$$\int_0^{2\pi} f(t)\psi_j(t)dt = 0, \quad j = 1, \ldots, l, \tag{4.167}$$

where l is the number of linearly independent solutions of the homogeneous problem (4.10), (4.165) and $\psi_1(t), \ldots, \psi_l(t)$ are some linearly independent continuous 2π-periodic functions dependent on $f(t)$.

In the preceding section we have obtained a condition for unique solvability of the general boundary value problem (4.10), (4.165) in two strata medium $x > 0$ (see Theorem 4.12). Here we are going to obtain simple conditions of Fredholmity for this problem in two strata medium $x > 0$.

First we consider the boundary condition of the form

$$H_y(0,t) = kE_z(0,t) + f(t), \tag{4.168}$$

where k is a real constant.

Recall that for $k \geq 0$ the problem (4.10), (4.168) in two strata medium $x > 0$ is uniquely solvable (see Theorem 4.7).

We investigate the Fredholmity of the problem (4.10), (4.168) in the medium described in section 4.3.

We have shown in section 4.3 that if $H_y(x,t)$ and $E_z(x,t)$ satisfy the problem (4.10), (4.168), then they will be defined by (4.16), (4.17), where

$$u_0(x) = A_0, \quad v_0(x) = 0, \tag{4.169}$$

and the functions $u_n(x)$ and $v_n(x)$ $(n = 1, 2, \ldots)$ are defined by (4.99), (4.102) and c_{n1}, c_{n2} and c_{n3} are constants satisfying the system of equations (4.109) – (4.111).

Excluding from (4.110) and (4.111) the constants c_{n1} and c_{n2} and substituting into equation (4.109) we get

$$Q_n c_{n3} = \frac{2i\lambda_n \mu_1 \mu_2 \mu_0 A_n}{n} \exp(-\mu_n x_1), \tag{4.170}$$

$$c_{n1} = \frac{\mu_2 \lambda_n + \mu_1 \nu_n}{2\lambda_n \mu_2} c_{n3} \exp(\mu_n x_1 - \lambda_n x_1), \tag{4.171}$$

$$c_{n2} = \frac{\mu_2 \lambda_n - \mu_1 \nu_n}{2\lambda_n \mu_2} c_{n3} \exp(\mu_n x_1 + \lambda_n x_1), \tag{4.172}$$

where Q_n is defined by (4.115).

Denote by Q the right-hand side of the relation (4.123):

$$\begin{aligned} Q &= \left(\mu_2\sqrt{\alpha_1} + \mu_1\sqrt{\alpha_2}\right) \exp\left(\frac{\beta_1\sqrt{\alpha_1}x_1}{\sqrt{2}}\right) \\ &\times \left(\mid k\mu_1\mu_0 + \sqrt{\alpha_1} \mid - \gamma \mid k\mu_1\mu_0 - \sqrt{\alpha_1} \mid\right), \end{aligned} \tag{4.173}$$

where

$$\gamma = \frac{\mid \mu_2\sqrt{\alpha_1} - \mu_1\sqrt{\alpha_2} \mid}{\mu_2\sqrt{\alpha_1} + \mu_1\sqrt{\alpha_2}} \exp(-\sqrt{2}\beta_1\sqrt{\alpha_1}x_1). \tag{4.174}$$

It is clear that $Q \neq 0$ if and only if

$$\mu_1\mu_0 k + \sqrt{\alpha_1} \neq \pm\gamma(\mu_1\mu_0 k - \sqrt{\alpha_1}). \tag{4.175}$$

From (4.175) we have

$$k \neq -\frac{(1+\gamma)\sqrt{\alpha_1}}{(1-\gamma)\mu_1\mu_0}, \quad -\frac{(1-\gamma)\sqrt{\alpha_1}}{(1+\gamma)\mu_1\mu_0}. \tag{4.176}$$

Let the condition (4.176) be satisfied. Then from (4.116) and (4.123) we have

$$\mid Q_n \mid \geq \frac{1}{2} \mid Q \mid > 0 \quad \text{for} \quad n \geq n_0, \tag{4.177}$$

where n_0 is a sufficiently large positive integer.

Using the inequality (4.177) we are going to prove the following

Theorem 4.14 *If the number k satisfies (4.176), then the problem (4.10), (4.168) in two strata medium $x > 0$ is Fredholmian.*

Proof. Let the condition (4.176) be satisfied. Then, according to (4.177) there exists only a finite number of positive integers $n_1, \ldots, n_l$ satisfying

$$Q_{n_j} = 0, \quad j = 1, \ldots, l, \tag{4.178}$$
$$Q_n \neq 0, \quad n \in N, \quad n \neq n_1, \ldots, n_l, \tag{4.179}$$

where N is the set of positive integers.

From (4.170), (4.178) and (4.179) we have

$$A_{n_j} = 0, \quad j = 1, \ldots, l, \tag{4.180}$$
$$c_{n_3} = \frac{2i\lambda_n \mu_1 \mu_2 \mu_0 A_n}{nQ_n} \exp(-\nu_n x_1), \quad n \neq n_j, \; j = 1, \ldots, l, \tag{4.181}$$

where $n \in N$ and $c_{n_1}, c_{n_2}, \ldots, c_{n_l}$ are arbitrary complex constants.

Thus, the conditions (4.180) are necessary for solvability of the problem (4.10), (4.168), where the constants A_n are defined by (2.28). Since the function $f(t)$ is real-valued, these conditions can be written as

$$\int_0^{2\pi} f(t) \cos n_j t dt = 0, \quad \int_0^{2\pi} f(t) \sin n_j t dt = 0 \quad j = 1, \ldots, l. \tag{4.182}$$

Let the condition (4.182) be satisfied. Then if $H_y(x,t)$ and $E_z(x,t)$ are solutions of the problem (4.10), (4.168) in two strata domain $x > 0$ then they will be defined by (4.16) and (4.17), where the functions $u_n(x)$ and $v_n(x)$ are defined by (4.99), (4.102) and (4.169), the constants c_{n_3} for $n \in N$, $n \neq n_j$, $j = 1, \ldots, l$ are defined by (4.181), while for $n = n_j$ $j = 1, \ldots, l$ they are arbitrary complex constants. The constants c_{n_1} and c_{n_2} $(n = 1, 2, \ldots)$ are defined by c_{n_3} by (4.171), (4.172).

Using the inequality (4.177) one can check that the functions $H_y(x,t)$ and $E_z(x,t)$ defined in the above mentioned way satisfy the considered problem for any complex constants $c_{n_j 3}$ $(j = 1, 2, \ldots, l)$. Substituting $f(t) = 0$ into this formula we get that the homogeneous problem has just $2l$ linearly independent solutions.

Theorem 4.14 is proved.

Now let us consider the boundary conditions of the form

$$\frac{\partial H_y(0,t)}{\partial x} = k \frac{\partial E_z(0,t)}{\partial x} - k_1 E_z(0,t) + k_2 H_y(0,t) + f(t), \tag{4.183}$$

or

$$\frac{\partial E_z(0,t)}{\partial x} = k_1 E_z(0,t) - k_2 H_y(0,t) + f(t), \tag{4.184}$$

where k, k_1, and k_2 are given real constants.

For $k \geq 0$, $k_1 \geq 0$, $k_2 > 0$ in section 4.3 it has been shown that the problems (4.10), (4.183) and (4.10), (4.184) in the two strata medium $x > 0$ are uniquely solvable (see Theorem 4.12).

One has the following two theorems.

Theorem 4.15 *If k satisfies the condition (4.176), then the problem (4.10), (4.183) is Fredholmian.*

Theorem 4.16 *The problem (4.10), (4.184) is Fredholmian.*

Theorems 4.15 and 4.16 can be proved in the same way as Theorem 4.14.

4.6 Harmonic Oscillations of Electromagnetic Waves in Multi-Strata Medium $x > 0$

Let the medium consist of m strata

$$x_{k-1} < x < x_k, \quad k = 1, 2, \ldots, m,$$

where $x_0 = 0$, $x_m = +\infty$. In this case the quantities ε, μ and σ determining the system of equations (4.10) are piecewise constant,

$$\varepsilon = \varepsilon_k, \quad \mu = \mu_k, \quad \sigma = \sigma_k \quad \textit{for} \quad x \in (x_{k-1}, x_k) \quad k = 1, \ldots, m, \tag{4.185}$$

where ε_k, σ_k and μ_k are positive constants.

A solution of the system (4.10) is searched in the same class as in the case of $m = 2$ (cf. section 4.4).

The aim of this section is to show that the results obtained in section 4.3 remains valid in the multi-strata medium $x > 0$.

For equation (4.10) we consider the boundary condition

$$H_y(x, 0) = kE_z(x, 0) + f(t), \tag{4.186}$$

where k is a non-negative constant and $f(t)$ is a triple continuously differentiable real 2π-periodic function.

One has the following

Theorem 4.17 *The problem (4.10), (4.186) in multi-strata medium $x > 0$ is uniquely solvable.*

Proof. Let $H_y(x, t)$ and $E_z(x, t)$ satisfy system (4.10) in multi-strata medium $x > 0$, and let $u_n(x)$ and $v_n(x)$ be the Fourier coefficients of these functions defined by (4.18). Then, for $x > 0$, $x \neq x_1, \ldots, x_m$ and $n = 0, 1, \ldots$ we have

$$\frac{dv_n(x)}{dx} = in\mu\mu_0 u_n(x), \quad \frac{du_n(x)}{dx} = (in\varepsilon\varepsilon_0 + \sigma)v_n(x). \tag{4.187}$$

Recall that in (4.187) the coefficients ε, μ and σ are piecewise constant and are defined by (4.185).

Since the functions $H_y(x,t)$ and $E_z(x,t)$ are continuous and bounded in the domain $x \geq 0$, $-\infty < t < +\infty$, it follows from (4.18) that the functions $u_n(x)$ and $v_n(x)$ $(n = 0, 1, \ldots)$ are also continuous and bounded for $x \geq 0$.

From (4.18) and (4.186) we have

$$u_n(0) = kv_n(0) + A_n, \quad n = 0, 1, \ldots, \tag{4.188}$$

where the constants A_n are defined by (4.28).

Resolving the problem (4.187), (4.188) for $n = 0$ in the class of bounded functions, we obtain

$$u_0(x) = A_0, \quad v_0(x) = 0, \quad x \geq 0. \tag{4.189}$$

From the first equation of (4.187) we have

$$u_n(x) = \frac{1}{in\mu\mu_0} \frac{dv_n(x)}{dx}, \quad n = 1, 2, \ldots. \tag{4.190}$$

Substituting $u_n(x)$ from (4.190) into the second equation of (4.187) and the boundary condition (4.188) we get

$$\frac{d^2 v_n(x)}{dx^2} = n\mu\mu_0(-n\varepsilon\varepsilon_0 + i\sigma)v_n(x), \ x > 0, \ x \neq x_1, \ldots, x_m, \tag{4.191}$$

$$v_n'(0) = kn\mu_1\mu_0 i v_n(0) + in\mu_1\mu_0 A_n. \tag{4.192}$$

We set

$$\lambda_{nk} = -\sqrt{n\mu_k\mu_0(-n\varepsilon_k\varepsilon_0 + i\sigma_k)}, \quad n = 1, 2, \ldots; \ k = 1, 2, \ldots, m. \tag{4.193}$$

Here we choose the branch of radical whose real part is positive.

From (4.190) it follows that the functions $v_n'(x)$ are also continuous on semi-axis $x > 0$.

The general solution of equation (4.191) on the interval (x_{m-1}, ∞) in the class of bounded function is defined by

$$v_n(x) = c_n \exp(\lambda_{nm} x), \tag{4.194}$$

where c_n is an arbitrary complex constant.

From (4.194) it follows that the functions $|v_n(x)|^2$ and $|v_n'(x)|^2$ are integrable on the interval (x_{m-1}, ∞).

Let us prove that the homogeneous problem (4.191), (4.192) (at $A_n = 0$) has only zero solution.

Indeed, let $v_n(x)$ be a solution of the homogeneous problem (4.191), (4.192) in the class of bounded functions. Then it is clear that

$$\int_0^\infty \left[\frac{d^2 v_n(x)}{dx^2} + n\mu\mu_0(n\varepsilon\varepsilon_0 - i\sigma)v_n(x)\right]\overline{v_n(x)}dx = 0, \tag{4.195}$$

where $\overline{v_n(x)}$ is the complex conjugate of $v_n(x)$.

Integrating by parts in (4.195) we get

$$v_n'(0)\overline{v_n(0)} + \int_0^\infty \left[| v'_n(x) |^2 + n\mu\mu_0(-n\varepsilon\varepsilon_0 + i\sigma) | v_n(x) |^2\right] dx = 0. \tag{4.196}$$

Substituting $v_n'(0)$ from (4.192) at $A_n = 0$ into (4.196) we get

$$in\mu_1\mu_0 k \mid v_n(0) \mid^2 + \int_0^\infty \mid v_n'(x) \mid^2 dx$$

$$+ \int_0^\infty n\mu\mu_0(-n\varepsilon\varepsilon_0 + i\sigma) \mid v_n(x) \mid^2 dx = 0. \tag{4.197}$$

Equalizing the imaginary parts of (4.197) we obtain

$$k\mu_1 \mid v_n(0) \mid^2 + \int_0^\infty \sigma\mu \mid v_n(x) \mid^2 dx = 0. \tag{4.198}$$

Since $k \geq 0$, $\sigma > 0$, and $\mu > 0$, from (4.198) we get

$$v_n(x) \equiv 0, \quad x \geq 0, \quad n = 1, 2, \ldots. \tag{4.199}$$

Substituting $v_n(x)$ from (4.199) into (4.190) we obtain

$$u_n(x) \equiv 0, \quad n = 1, 2, \ldots. \tag{4.200}$$

Hence the homogeneous problem (4.10), (4.186) (at $f \equiv 0$) has only zero solution.

Now we prove the existence of a solution of the non-homogeneous problem (4.10), (4.186) in the class of bounded functions. To this end we construct a particular solution $w_n(x)$ of equation (4.191) as follows

$$w_n(x) = \exp(\lambda_{mn}x), \quad x \in (x_{m-1}, \infty), \tag{4.201}$$

$$w_n(x) = c_{nk}\exp(\lambda_{nk}x) + d_{nk}\exp(-\lambda_{nk}x), \tag{4.202}$$

$$x \in [x_{k-1}, x_k), \; k = 1, 2, \ldots, m-1$$

where the constants c_{nk} and d_{nk} for $k = 1, 2, \ldots, m-1$ are defined by recurrence relations

$$\begin{aligned} c_{nk} \exp(\lambda_{nk} x_k) + d_{nk} \exp(-\lambda_{nk} x_k) &= w_n(x_k), & (4.203) \\ \lambda_{nk} [c_{nk} \exp(\lambda_{nk} x_k) - d_{nk} \exp(-\lambda_{nk} x_k)] &= w_n'(x_k). & (4.204) \end{aligned}$$

Resolving the system (4.203), (4.204) with respect to the constants c_{nk} and d_{nk} for $k = 1, 2, \ldots, m-1$ we get

$$\begin{aligned} c_{nk} &= \frac{1}{2}[w_n(x_k) + \frac{1}{\lambda_{nk}} w_n'(x_k)] \exp(-\lambda_{nk} x_k), & (4.205) \\ d_{nk} &= \frac{1}{2}[w_n(x_k) - \frac{1}{\lambda_{nk}} w_n'(x_k)] \exp(\lambda_{nk} x_k). & (4.206) \end{aligned}$$

Substituting $w_n(x)$ from (4.201) into (4.205) and (4.206) at $k = m-1$ we get $c_{n,m-1}$ and $d_{n,m-1}$. Substituting the obtained values of $c_{n,m-1}$ and $d_{n,m-1}$ into (4.202) we determine the function $w_n(x)$ on the interval $[x_{m-2}, x_{m-1}]$. In the same way we can construct a particular solution $w_n(x)$ of equation (4.291). From (4.201), (4.202), (4.205) and (4.206) it follows that $w_n(x)$ is continuously differentiable and bounded for $x \geq 0$.

Since the homogeneous problem (4.291), (4.292) (at $A_n = 0$) in the class of bounded functions has only zero solution, then

$$w_n'(0) - ik\mu_1\mu_0 n w_n(0) \neq 0. \quad (4.207)$$

Indeed, if this is not the case, $w_n(x)$ will be a non-zero solution of the homogeneous problem (4.291), (4.292) (at $A_n = 0$) in the class of bounded functions, but this is impossible.

We seek a solution of the non-homogeneous problem (4.291), (4.292) as

$$v_n(x) = c_n w_n(x), \quad (4.208)$$

where the constant c_n should be defined.

Substituting $v_n(x)$ from (4.208) into the boundary condition (4.292) we get

$$c_n = \frac{inA_n\mu_1\mu_0}{w_n'(0) - ink\mu_1\mu_0 w_n(0)}. \quad (4.209)$$

From (4.190), (4.208) and (4.209) we have

$$\begin{aligned} v_n(x) &= \frac{in\mu_1\mu_0 A_n w_n(x)}{w_n'(0) - ink\mu_1\mu_0 w_n(0)}, \\ u_n(x) &= \frac{A_n\mu_1 w_n'(x)}{\mu(w_n'(0) - ink\mu_1\mu_0 w_n(0))}. & (4.210) \end{aligned}$$

Substituting $u_n(x)$ and $v_n(x)$ from (4.189) and (4.210) into (4.16) and (4.17) we get

$$H_y(x,t) = A_0 + 2\sum_{n=1}^{\infty} Re\left[\frac{A_n\mu_1 w_n'(x)}{\mu(w_n'(0) - ink\mu_1\mu_0 w_n(0))}\exp(int)\right], \quad (4.211)$$

$$E_z(x,t) = 2\sum_{k=1}^{\infty} Re\left[\frac{inA_n\mu_1\mu_0 w_n(x)}{w_n'(0) - ink\mu_1\mu_0 w_n(0)}\exp(int)\right]. \quad (4.212)$$

Let us show that the functions $H_y(x,t)$ and $E_z(x,t)$ defined by (4.211) and (4.212) satisfy the problem (4.10), (4.186).

Consider two cases.

Case 1. Let $f(t)$ be a trigonometric polynomial

$$f(t) = A_0 + 2\sum_{n=1}^{N} Re[A_n \exp(int)], \quad (4.213)$$

where A_0 is real, while A_n are complex constants. Then in (4.211) and (4.212) we have $A_n = 0$ at $n \geq N+1$. It is easy to check that the functions defined by (4.211), (4.212) are solutions of the problem (4.10), (4.186).

Case 2. Let $f(t)$ be real triple continuously differentiable 2π-periodic function. Then it admits the Fourier expansion

$$f(t) = A_0 + 2\sum_{n=1}^{\infty} Re[A_n \exp(int)], \quad (4.214)$$

where the coefficients A_n are defined by (4.28) and satisfy

$$| A_n | \leq \frac{c}{n^3}, \quad n = 1, 2, \ldots, \quad (4.215)$$

where the constant c does not depend on n.

We prove that in this case the series (4.211) and (4.212) converge uniformly and admit differentiation term by term on x and t.

From (4.22) and (4.193) we have

$$| \exp(\lambda_{nk}x) | \leq 1, \quad x \geq 0, \quad k = 1, \ldots, m, \quad (4.216)$$

$$| \exp(\lambda_{nm}x) | \geq \exp(-\frac{\beta_m\sqrt{\alpha_m}x}{2}), \quad x \geq x_{m-1}, \quad (4.217)$$

$$| \exp(-\lambda_{nk}x) | \leq c_0, \quad 0 \leq x \leq x_{m-2}, \quad k = 1, \ldots, m-1, \quad (4.218)$$

where c_0 is a positive constant independent of n and

$$\alpha_m = \varepsilon_0\mu_0\varepsilon_m\mu_m, \quad \beta_m = \frac{\sigma_m}{\varepsilon_0\varepsilon_m}.$$

From (4.201), (4.202), (4.205), (4.206) and (4.216) – (4.218), the estimates follow:

$$\left| \frac{d^j w_n(x)}{dx^j} \right| \leq N_0 n^j, \quad x \geq 0, \ j = 0, 1, 2; \ n = 1, 2, \ldots, \tag{4.219}$$

$$\int\limits_{x_m}^{\infty} | \ w_n(x) \ |^2 \ dx \geq N_1, \quad n = 1, 2, \ldots, \tag{4.220}$$

where N_0 and N_1 are some positive constants independent of n.

Since $w_n(x)$ is a bounded solution of equation (4.191) it satisfies the equality (4.196). Substituting $v_n(x) = w_n(x)$ into (4.196) and equalizing the imaginary parts to zero, we get

$$Im[w_n'(0)\overline{w_n(0)}] + \int\limits_0^{\infty} n\mu\mu_0\sigma \ | \ w_n(x) \ |^2 \ dx = 0. \tag{4.221}$$

The equality (4.221) can be written as follows

$$Im[(ink\mu_1\mu_0 w_n(0) - w_n'(0))\overline{w_n(0)}]$$
$$= \int\limits_0^{\infty} n\mu\mu_0\sigma \ | \ w_n(x) \ |^2 \ dx + nk\mu_1\mu_0 \ | \ w_n(0) \ |^2 . \tag{4.222}$$

Since $\sigma > 0, \ k \geq 0$ and $\mu > 0$, from (4.220) and (4.222) we have

$$Im[(ink\mu_1\mu_0 w_n(0) - w_n'(0))\overline{w_n(0)}] \geq n\mu_0\mu_m\sigma_m N_1. \tag{4.223}$$

It is clear that

$$| \ ink\mu_1\mu_0 w_n(0) - w_n'(0) \ || \ w_n(0) \ | \geq Im[(inkw_n(0) - w_n'(0))\overline{w_n(0)}]. \tag{4.224}$$

From (4.219) for $j = 0$ and $x = 0$ we have

$$| \ ink\mu_1\mu_0 w_n(0) - w_n'(0) \ || \ w_n(0) \ | \leq | \ ink\mu_1\mu_0 w_n(0) - w_n'(0) \ | \ N_0. \tag{4.225}$$

From the inequalities (4.223) – (4.225) we deduce

$$| \ ink\mu_1\mu_0 w_n(0) - w_n'(0) \ | \ N_0 \geq n\mu_0\mu_m\sigma_m N_1. \tag{4.226}$$

Hence

$$| \ w_n'(0) - ink\mu_1\mu_0 w_n(0) \ | \geq \frac{n\mu_0\mu_m\sigma_m N_1}{N_0}. \tag{4.227}$$

From (4.215), (4.219) and (4.227) it follows that the series (4.211) and (4.212) converge uniformly and admit term by term differentiation on x and t.

Substituting $H_y(x,t)$ and $E_z(x,t)$ from (4.211), (4.212) into the system of equations (4.10) and the boundary condition (4.186) we conclude that these functions satisfy the problem (4.10), (4.186). So the problem (4.10), (4.186) is uniquely solvable.

Theorem 4.17 is proved.

Now let us consider a more general boundary condition of the form

$$H_y(0,t) = k_1\frac{\partial H_y(0,t)}{\partial x} - k_2\frac{\partial E_z(0,t)}{\partial x} + k_3E_z(0,t) + f(t), \qquad (4.228)$$

where k_1, k_2 and k_3 are non-negative real constants and $f(t)$ is the same function as in (4.186).

Denote

$$l_n = \frac{w_n'(0)}{in\mu_1\mu_0} - k_1(in\varepsilon_1\varepsilon_0 + \sigma_1)w_n(0) + k_2w_n'(0) - k_3w_n(0), \qquad (4.229)$$

where $w_n(x)$ is the function from (4.201) and (4.202).

Using the arguments of the proof of (4.207), it is easy to show that

$$l_n \neq 0, \quad n = 1, 2, \ldots. \qquad (4.230)$$

One has the following

Theorem 4.18 *The problem (4.10), (4.228) in multi-strata medium is uniquely solvable and the solution is defined by*

$$H_y(x,t) = A_0 + 2\sum_{n=1}^{\infty} Re\left[\frac{A_nw_n(x)\exp(int)}{in\mu_1\mu_0l_n}\right], \qquad (4.231)$$

$$E_z(x,t) = 2\sum_{n=1}^{\infty} Re\left[\frac{A_nw_n'(x)\exp(int)}{l_n}\right]. \qquad (4.232)$$

Proof is similar to that of Theorem 4.17.

Now we consider the boundary condition of the form

$$k_1\frac{\partial H_y(0,t)}{\partial x} - k_2\frac{\partial E_z(0,t)}{\partial x} + k_3E_z(0,t) = f(t), \qquad (4.233)$$

where k_1, k_2 and k_3 are real constants satisfying

$$k_1^2 + k_2^2 + k_3^2 \neq 0, \quad k_1k_2 \geq 0.$$

It is easy to check that $H_y(x,t) = c$, $E_z(x,t) = 0$ is a solution of the homogeneous problem (4.10), (4.233), where c is an arbitrary constant. To determine the constant c we impose on $H_y(0,t)$ a supplementary condition

$$\frac{1}{2\pi}\int_0^{2\pi} H_y(0,\tau)d\tau = \delta_0, \qquad (4.234)$$

where δ_0 is a given real number.

Resolving the system of equations (4.187) for $n = 0$ in the class of bounded functions we get

$$v_0(x) \equiv 0, \quad u_0(x) = const, \quad x \geq 0. \tag{4.235}$$

From (4.235) we obtain

$$k_1 u_0'(0) - k_2 v_0'(0) + k_3 v_0(0) = 0. \tag{4.236}$$

Substituting $u_n(x)$ and $v_n(x)$ at $n = 0$ from (4.18) into (4.236) we get

$$\int_0^{2\pi} [k_1 \frac{\partial H_y(0,t)}{\partial x} - k_2 \frac{\partial E_z(0,t)}{\partial x} + k_3 E_z(0,t)] dt = 0. \tag{4.237}$$

From (4.233) and (4.237) we have

$$\int_0^{2\pi} f(t) dt = 0. \tag{4.238}$$

Hence the condition (4.238) is necessary for the solvability of the problem (4.10), (4.233). Below we assume that (4.238) is satisfied.

One has the following

Theorem 4.19 *The problem (4.10), (4.233), (4.234) is uniquely solvable and the solution is defined by*

$$B_y(x,t) = \delta_0 + 2 \sum_{n=1}^{\infty} Re[\frac{A_n w_n'(x) \exp(int)}{in\mu_1\mu_0 q_n}],$$

$$E_z(x,t) = 2 \sum_{n=1}^{\infty} Re[\frac{A_n w_n(x) \exp(int)}{q_n}],$$

where

$$q_n = k_1(\sigma_1 + in\varepsilon_1\varepsilon_0) w_n(0) - k_2 w_n'(0) + k_3 w_n(0).$$

Proof is similar to that of Theorem 4.17.

4.7 Harmonic Oscillations of Electromagnetic Waves in Non-Homogeneous Media

In this section we consider electromagnetic fields, for which $\mu = \mu(x)$, $\varepsilon = \varepsilon(x)$ and $\sigma = \sigma(x)$ are continuously differentiable functions on x and

$$\mu(x) > 0, \quad \varepsilon(x) > 0, \quad \sigma(x) > 0 \quad \text{for} \quad x \geq 0,$$

$$\mu(x) = \mu_2, \quad \varepsilon(x) = \varepsilon_2, \quad \sigma(x) = \sigma_2 \quad \text{at} \quad x \geq x_1.$$

Here μ_2, ε_2 and σ_2 are positive constants and x_1 is a sufficiently large positive number.

We investigate an electromagnetic field depending only on x and t, which is 2π-periodic on t. Note that in this case the system of equations (4.7) – (4.10) and (4.187) remain the same.

For the system (4.10) we take the boundary condition

$$H_y(0,t) \;=\; kE_z(0,t) + f(t), \tag{4.239}$$

where k is a non-negative constant and $f(t)$ is a trigonometric polynomial

$$f(t) \;=\; A_0 + 2Re\sum_{k=1}^{N} A_n e^{int}, \tag{4.240}$$

where A_0 is real and $A_1, \ldots, A_n$ are complex constants.

As it has been mentioned in section 4.2 the solutions $H_y(x,t)$ and $E_z(x,t)$ of (4.10) admit the expansions (4.16) and (4.17), where $u_n(x)$ and $v_n(x)$ for $x > 0$ satisfy the following system

$$\frac{dv_n(x)}{dx} \;=\; in\mu(x)\mu_0 u_n(x), \qquad \frac{du_n(x)}{dx} \;=\; (in\varepsilon(x)\varepsilon_0 + \sigma(x))v_n(x). \tag{4.241}$$

From (4.18) and (4.239) we get

$$u_n(0) \;=\; kv_n(0) + A_n, \quad n \;=\; 0, 1, \ldots, \tag{4.242}$$

where A_n for $n = 0, \ldots, N$ are the coefficients from the expansion (4.240) and $A_n = 0$ for $n = N+1, N+2, \ldots$.

Since $H_y(x,t)$ and $E_z(x,t)$ are searched in the class of continuously differentiable and bounded in R_2^+ functions, it follows from (4.11) that $u_n(x)$ and $v_n(x)$ are also continuously differentiable and bounded on $x \geq 0$.

Let $n = 0$. A solution of the problem (4.241), (4.242) is defined by

$$u_0(x) \;=\; A_0, \quad v(0) \;=\; 0, \quad x \geq 0. \tag{4.243}$$

Let $n \geq 1$. Then from the first equation of the system (4.241) we have

$$u_n(x) \;=\; \frac{1}{in\mu(x)\mu_0} v_n'(x), \quad x > 0, \quad n \;=\; 1, 2, \ldots. \tag{4.244}$$

Substituting $u_n(x)$ from (4.244) into the second equation of the system (4.241) and the boundary condition (4.242) we get

$$\frac{d}{dx}\left(\frac{v_n'(x)}{\mu(x)}\right) + (n^2\varepsilon(x)\varepsilon_0 - in\sigma(x))\mu_0 v_n(x) \;=\; 0, \quad x > 0, \tag{4.245}$$

$$v_n'(0) \;=\; in\mu(0)\mu_0 k v_n(0) + in\mu(0)\mu_0 A_n. \tag{4.246}$$

In the case, where ε, μ and σ are piecewise constant, in section 4.5 it has been shown that the problem (4.245), (4.246) for $A_n = 0$ in the class of bounded functions has only zero solution. In a similar way one can prove that this statement remains valid in this case.

Now we turn to the construction of a solution of the problem (4.245), (4.246). We set

$$w_n(x) = \exp(-\lambda_{n2}x), \quad x \geq x_1,$$

where

$$\lambda_{n2} = \sqrt{(-n^2\varepsilon_2\varepsilon_0 + i\sigma_2 n)\mu_2\mu_0}.$$

Here we choose the branch of radical whose real part is positive.

Let $w_n(x)$ at $0 \leq x \leq x_1$ be a solution of equation (4.245) with Cauchy boundary condition

$$w_n(x_1) = \exp(-\lambda_{n2}x_1), \quad w_n'(x_1) = -\lambda_{n2}\exp(-\lambda_{n2}x_1).$$

It is clear that

$$w_n'(0) - in\mu(0)\mu_0 k w_n(0) \neq 0,$$

otherwise $w_n(x)$ will be a non-zero solution of the problem (4.245), (4.246) for $A_n = 0$ in the class of bounded functions, which is impossible.

It is easy to check that the function

$$v_n(x) = \frac{inA_n\mu(0)\mu_0 w_n(x)}{w_n'(0) - ikn\mu(0)\mu_0} \tag{4.247}$$

is a solution of the problem (4.245), (4.246) in the class of bounded functions.

Substituting $v_n(x)$ from (4.247) into (4.244) we obtain

$$u_n(x) = \frac{A_n\mu(0)w_n'(x)}{\mu(x)(w_n'(0) - ink\mu(0)\mu_0)}. \tag{4.248}$$

Substituting $v_n(x)$ and $u_n(x)$ from (4.247) and (4.248) into (4.16) and (4.17) we get a solution of the problem (4.10), (4.239). Hence, the problem (4.10), (4.239) is uniquely solvable.

Note that equation (4.10) with the boundary conditions (4.228) or (4.233) can be investigated similarly.

Chapter 5

Calculation of Capacitances of Cylindrical and Spherical Capacitors

5.1 Introduction

There is a great number of works in scientific literature devoted to the calculations of capacitances of different kinds of capacitors (cf., e.g., [18] - [24]). In these works the approximate formulae are mostly obtained and, in general, the exactness of these formulae is not investigated. The exact formulae are obtained for cylindrical capacitors with cross-sections bounded by circumferences or by confocal ellipses.

The main difference between the present chapter and the above mentioned papers is as follows:
1) We obtain exact formulae of capacitances for a wide class of cylindrical capacitors with cross-sections bounded by analytic curves;
2) We obtain approximate formulae of capacitances for cylindrical capacitors with arbitrary cross-sections and estimate the absolute and relative errors of these formulae;
3) We resolve some problems of choice of cylindrical capacitors with given parameters;
4) We describe some classes of mappings of cross-sections of cylindrical capacitors preserving the invariance of capacitance. These mappings permit us to choose the capacitors with given parameters.
5) We describe a new method of determining of capacitances for cylindrical capacitors.

5.2 Invariance of Capacitances under Conformal Mappings

Because this chapter is mainly for engineers, we recall some known definitions and facts from the theory of analytic and harmonic functions.

Definition 5.1 A capacitor K bounded by two cylindrical surfaces is called cylindrical capacitor.

In the usual way we identify the plane OXY with the complex plane $\mathbf{C} : ((x, y) \leftrightarrow z = x + iy)$.

Let D be a domain in the complex plane and let $f(z)$ be a complex valued function in D. The function $f(z)$ is said to be analytic in D if it is differentiable at any point z with respect to the complex variable $z = x + iy \in D$.

Let D_1 and D_2 be biconnected finite domains in the complex plane with sufficiently smooth boundaries. Let Γ_j be the exterior and γ_j be the interior boundaries of the domains D_j $(j = 1, 2)$.

Definition 5.2 A function $f(z)$ is said to be a conformal mapping from D_1 onto D_2 if the following conditions are satisfied:

1) $f(z)$ is analytic in D_1,

2) $f(z)$ is one-to-one mapping from D_1, Γ_1 and γ_1 onto D_2, Γ_2 and γ_2, respectively.

3) $f'(z) \neq 0$, for $z \in D_1$.

In the case where D_1 and D_2 are simple connected finite domains, by Riemann's theorem, there always exists a conformal mapping from D_1 onto D_2. In the case of biconnected domains D_1 and D_2 this assertion breaks down. If, for example, D_1 and D_2 are the rings

$$r_1^2 < x^2 + y^2 < R_1^2 \quad \text{and} \quad r_2^2 < x^2 + y^2 < R_2^2,$$

such a mapping exists if and only if

$$\frac{R_1}{r_1} = \frac{R_2}{r_2}. \tag{5.1}$$

Definition 5.3 Two domains D_1 and D_2 are said to be equal within conformal mapping accuracy, if there exists a conformal mapping from D_1 onto D_2.

Theorem 5.1 *The capacitances of two cylindrical capacitors K_1 and K_2 are equal if and only if their cross-sections D_1 and D_2 are equal within conformal mapping accuracy.*

The sufficiency part of Theorem 5.1 can be found (without proof) in [19] (chap. 2, section 6). Here we present the complete proof of this theorem.

Proof of Theorem 5.1. Let the cross-sections D_1 and D_2 of capacitors K_1 and K_2 be equal within conformal mapping accuracy. This means that there exists an analytic in D_1 function $f(z)$, which maps conformally D_1 onto D_2.

We are going to prove that the capacitances of these capacitors are equal. To this end let us recall the formulae of the capacitances C_1 and C_2. Let $u_1(x,y)$ and $u_2(x,y)$ be harmonic functions in D_1 and D_2, respectively and satisfy the following boundary conditions

$$u_1(x,y) = 1, \quad (x,y) \in \Gamma_1, \tag{5.2}$$

$$u_1(x,y) = 0, \quad (x,y) \in \gamma_1, \tag{5.3}$$

$$u_2(x,y) = 1, \quad (x,y) \in \Gamma_2, \tag{5.4}$$

$$u_2(x,y) = 0, \quad (x,y) \in \gamma_2. \tag{5.5}$$

It is known (see [4], p.102), that the capacitances C_1 and C_2 of the cylindrical capacitors K_1 and K_2 per unit length are defined by

$$C_1 = \varepsilon\varepsilon_0 \int\limits_{\Gamma_1} \frac{\partial u_1(x,y)}{\partial N_1} ds_1, \tag{5.6}$$

$$C_2 = \varepsilon\varepsilon_0 \int\limits_{\Gamma_2} \frac{\partial u_2(\zeta,\eta)}{\partial N_2} ds_2, \tag{5.7}$$

where N_j $(j = 1,2)$ is the outer normal to the boundary Γ_j at the point of integration, ε_0 is the electric constant, ε is dielectric permitivity, ds_1 and ds_2 are the elements of the arcs of the curves Γ_1 and Γ_2, respectively.

Let $f(z) = \alpha(x,y) + i\beta(x,y)$, where $\alpha(x,y)$ and $\beta(x,y)$ are the real and imaginary parts of an analytic function $f(z)$. Since $f(z)$ maps conformally D_1 onto D_2 and $u_2(\zeta,\eta)$ is harmonic in D_2 function, the function

$$u(x,y) \equiv u_2(\alpha(x,y), \beta(x,y)) \tag{5.8}$$

is harmonic in D_1 (see [14]).

From (5.4) and (5.5) we have

$$u(x,y) = 1, \quad (x,y) \in \Gamma_1, \tag{5.9}$$

$$u(x,y) = 0, \quad (x,y) \in \gamma_1. \tag{5.10}$$

Thus, the harmonic in D_1 functions $u_1(x,y)$ and $u(x,y)$ coincide on the boundary of D_1, therefore (see [14]) they also coincide in D_1, i.e.

$$u_1(x,y) = u_2(\alpha(x,y), \beta(x,y)). \tag{5.11}$$

Since the function $\zeta = f(z)$ maps conformally the domain D_1 onto the domain D_2, it follows from (5.11) that

$$\frac{\partial u_1(x,y)}{\partial N_1} = \frac{\partial u_2(\xi,\eta)}{\partial N_2} \mid f'(z) \mid, \quad (x,y) \in \Gamma_1, \tag{5.12}$$

where $\xi = \alpha(x,y)$ and $\eta = \beta(x,y)$.

For this mapping we have

$$ds_2 = ds_1 \mid f'(z) \mid, \tag{5.13}$$

where ds_1 and ds_2 are the elements of arcs of curves Γ_1 and Γ_2, respectively.

Changing the variable

$$\xi + i\eta = f(z) \tag{5.14}$$

in the integral (5.7) and using the relations (5.12) and (5.13) we get

$$C_2 = \varepsilon\varepsilon_0 \int\limits_{\Gamma_2} \frac{\partial u_2(\xi,\eta)}{\partial N_2} ds_2 = \varepsilon\varepsilon_0 \int\limits_{\Gamma_1} \frac{\partial u_1(x,y)}{\partial N_1} ds_1 = C_1. \tag{5.15}$$

Thus, if the cross-sections of two cylindrical capacitors K_1 and K_2 coincide within conformal mapping accuracy, then their capacitances are equal.

Now we prove the inverse assertion. Let the capacitances of two cylindrical capacitors be equal. We prove that their cross-sections D_1 and D_2 coincide within conformal mapping accuracy.

It is known (see [14]), that there exist rings $r_1 \leq\mid z \mid\leq R_1$ and $r_2 \leq\mid z \mid\leq R_2$ coinciding with D_1 and D_2 within conformal mapping accuracy. Let K_1^* and K_2^* be cylindrical capacitors with cross-sections $r_1 \leq\mid z \mid\leq R_1$ and $r_2 \leq\mid z \mid\leq R_2$, respectively.

It is known (see [20], p. 64), that the capacitances C_1^* and C_2^* of capacitors K_1^* and K_2^* are defined by

$$C_1^* = 2\pi\varepsilon\varepsilon_0 \cdot \left(ln\frac{R_1}{r_1}\right)^{-1}, \quad C_2^* = 2\pi\varepsilon\varepsilon_0 \cdot \left(ln\frac{R_2}{r_2}\right)^{-1}. \tag{5.16}$$

We have

$$C_1 = C_1^*, \quad C_2 = C_2^*. \tag{5.17}$$

Since by assumption $C_1 = C_2$, it follows from (5.16) and (5.17) that

$$\frac{R_1}{r_1} = \frac{R_2}{r_2}. \tag{5.18}$$

From (5.18) the function

$$\zeta = R_2 R_1^{-1} z$$

maps conformally the ring $r_1 \leq |z| \leq R_1$ onto the ring $r_2 \leq |z| \leq R_2$, i.e. these two rings are equal within conformal mapping accuracy. Hence, the following domains are also equal within conformal mapping accuracy:

a) the domain D_1 and the ring $r_1 < |z| < R_1$,

b) the rings $r_1 < |z| < R_1$ and $r_2 < |z| < R_2$,

c) the ring $r_2 < |z| < R_2$ and the domain D_2.

Therefore the domains D_1 and D_2 are equal within conformal mapping accuracy.

Theorem 5.1 is proved.

Thus, using the conformal mappings of the ring $r < |z - z_0| < R$ we obtain all the possible cylindrical capacitors whose cross-sections coincide with this ring within conformal mapping accuracy. The capacitance of any such capacitor is defined by

$$C = 2\pi\varepsilon\varepsilon_0 \cdot \left(ln \frac{R}{r} \right)^{-1} . \tag{5.19}$$

5.3 Formulae of Capacitances with Cross-Sections Bounded by Analytic Curves

Using Theorem 5.1 and formula (5.19) in this section we obtain exact formulae for capacitances of the cylindrical capacitors bounded by analytic curves.

Here we calculate the capacitance per unit length of cylindrical capacitor.

1. First let the cross-section of cylindrical capacitor K be bounded by two arbitrary circumferences. The capacitance of these capacitors are defined by equipotential surfaces of two-point charges (cf. [21], p. 92).
We are going to obtain this formula in the general way, using our Theorem 5.1 and formula (5.19).

Let the cross-section D_1 of capacitor K_1 be bounded by circumferences

$$\Gamma_1 : \quad (\xi - x_0)^2 + \eta^2 = R^2, \tag{5.20}$$

$$\Gamma_2 : \quad (\xi - x_0 + d)^2 + \eta^2 = r^2, \tag{5.21}$$

where $x_0 > R$, $d > 0$ and $r + d < R$.

Consider the mapping

$$z = \zeta^{-1}, \quad (z = x + iy, \quad \zeta = \xi + i\eta). \tag{5.22}$$

The function (5.22) maps conformally D_1 onto G_1, bounded by two circumferences γ_1 and γ_2. The centers of these circumferences are at the points $(x_1, 0)$ and $(x_2, 0)$ $(x_1, x_2 > 0)$ with radii R_1 and R_2 defined by

$$R_1 = \frac{1}{2} \left[\frac{1}{x_0 - R} - \frac{1}{x_0 + R} \right], \tag{5.23}$$

$$R_2 = \frac{1}{2}\left[\frac{1}{x_0 - d - r} - \frac{1}{x_0 - d + r}\right], \tag{5.24}$$

$$x_1 = \frac{1}{2}\left[\frac{1}{x_0 - R} + \frac{1}{x_0 + R}\right], \tag{5.25}$$

$$x_2 = \frac{1}{2}\left[\frac{1}{x_0 - d - r} + \frac{1}{x_0 - d + r}\right]. \tag{5.26}$$

We choose x_0 such that the centers of circumferences γ_1 and γ_2 coincide, i.e.

$$\frac{1}{x_0 - R} + \frac{1}{x_0 + R} = \frac{1}{x_0 - d - r} + \frac{1}{x_0 - d + r}. \tag{5.27}$$

Resolving equation (5.27) with respect to x_0 we get

$$x_0 = \frac{R^2 - r^2 + d^2 + \sqrt{(R^2 - r^2 + d^2)^2 - 4R^2 d^2}}{2d}. \tag{5.28}$$

The condition $R - d - r > 0$ implies $x_0 > R$. Let x_0 be defined by (5.28), then $z = \zeta^{-1}$ maps conformally D_1 onto the ring $R_2 <| z - x_0 |< R_1$.

So, according to Theorem 5.1 and formula (5.19), the capacitance C is defined by

$$C = 2\pi\varepsilon\varepsilon_0 \cdot \left(ln\frac{R_1}{R_2}\right)^{-1}. \tag{5.29}$$

Now let the cross-section D of the capacitor K be bounded by circumferences L_1 and L_2:

$$L_1 : \quad x^2 + y^2 = R^2, \tag{5.30}$$

$$L_2 : \quad (x + d)^2 + y^2 = r^2, \tag{5.31}$$

where $0 < d < R$ and $r + d < R$.

It is clear that the capacitances of the capacitors K and K_1 coincide, since these two capacitors are geometrically identical.

So we get the following

Theorem 5.2 *Let the cross-section of a capacitor K be bounded by circumferences (5.30) and (5.31). Then the capacitance of K is defined by*

$$C = 2\pi\varepsilon\varepsilon_0 \cdot \left(ln\frac{R_1}{R_2}\right)^{-1}, \tag{5.32}$$

where

$$R_1 = R \cdot (x_0^2 - R^2)^{-1}, \tag{5.33}$$

$$R_2 = \frac{r}{(x_0 - d)^2 - r^2}, \tag{5.34}$$

$$x_0 = \frac{R^2 - r^2 + d^2 + \sqrt{(R^2 - r^2 + d^2)^2 - 4R^2 d^2}}{2d}. \tag{5.35}$$

2. Now let the cross-section D of a cylindrical capacitor K be bounded by the confocal ellipses

$$\Gamma_1 : \quad \frac{\xi^2}{A^2} + \frac{\eta^2}{B^2} = 1, \tag{5.36}$$

$$\Gamma_2 : \quad \frac{\xi^2}{a^2} + \frac{\eta^2}{b^2} = 1, \tag{5.37}$$

where the axes of these ellipses satisfy the conditions

$$A > B, \quad a > b, \quad A > a, \quad B > b, \tag{5.38}$$

$$A^2 - B^2 = a^2 - b^2. \tag{5.39}$$

The condition (5.39) stipulate the coincidence of foci of ellipses Γ_1 and Γ_2. Consider the following mapping

$$\zeta = \frac{1}{2}(z + \frac{\alpha^2}{z}), \tag{5.40}$$

where

$$z = x + iy, \quad \zeta = \xi + i\eta, \quad \alpha = \sqrt{A^2 - B^2}. \tag{5.41}$$

It is known (see [14]), that the function (5.40) maps conformally the domain $x^2 + y^2 > \alpha^2$ onto the complement of segment $[-\alpha, \alpha]$.

Let us consider the circumference

$$x^2 + y^2 = \rho^2, \tag{5.42}$$

where $\rho > \alpha$. We are going to find two values $\rho = \rho_1$ and $\rho = \rho_2$ for which the images of the circumferences $x^2 + y^2 = \rho_i^2$ coincide with contours Γ_j $(j = 1, 2)$.

The parametric equation of circumference (5.42) in a complex form looks as follows

$$z = \rho \cos\psi + i\rho \sin\psi = \rho \exp(i\psi), \quad 0 \leq \psi \leq 2\pi. \tag{5.43}$$

Substituting (5.43) into (5.40) we get

$$\zeta = \frac{1}{2}(\rho e^{i\psi} + \frac{\alpha^2}{\rho} e^{-i\psi}). \tag{5.44}$$

Equalizing the real and imaginary parts in (5.44), we get

$$\xi = \frac{1}{2}(\rho + \frac{\alpha^2}{\rho}) \cos\psi, \quad \eta = \frac{1}{2}(\rho - \frac{\alpha^2}{\rho}) \sin\psi. \tag{5.45}$$

Equation (5.45) is a parametric equation of the image of the circumference (5.42) under the mapping (5.40). On the other hand, it is parametric equation of an ellipse with axes A_0 and B_0 equal to

$$A_0 = \frac{1}{2}(\rho + \frac{\alpha^2}{\rho}), \quad B_0 = \frac{1}{2}(\rho - \frac{\alpha^2}{\rho}). \tag{5.46}$$

Now we choose positive constants ρ_1 and ρ_2 such that

$$\frac{1}{2}(\rho_1 + \frac{\alpha^2}{\rho_1}) = A, \quad \frac{1}{2}(\rho_1 - \frac{\alpha^2}{\rho_1}) = B, \tag{5.47}$$

$$\frac{1}{2}(\rho_2 + \frac{\alpha^2}{\rho_2}) = a, \quad \frac{1}{2}(\rho_2 - \frac{\alpha^2}{\rho_2}) = b. \tag{5.48}$$

We are going to prove that the system of equations (5.47) and (5.48), under the conditions (5.38) and (5.39), is solvable with respect to ρ_1 and ρ_2, and the solutions satisfy the condition $\alpha < \rho_2 < \rho_1$.

The first equation of the system (5.47) can be written as

$$\rho_1^2 - 2A\rho_1 + \alpha^2 = 0. \tag{5.49}$$

Equation (5.49) has two real solutions, the greater of which is defined by

$$\rho_1^2 = A + \sqrt{A^2 - \alpha^2}. \tag{5.50}$$

Substituting α from (5.41) into (5.50) we get

$$\rho_1 = A + B. \tag{5.51}$$

Using the equality $\alpha^2 = A^2 - B^2$, it is easy to check that $\rho_1 = A + B$ is a solution of the second equation in (5.47).

In a similar way we prove that

$$\rho_2 = a + b \tag{5.52}$$

satisfies the system (5.48). Since $\alpha = \sqrt{A^2 - B^2} = \sqrt{a^2 - b^2}$, we have

$$\alpha < a. \tag{5.53}$$

The inequality $\alpha < \rho_2 < \rho_1$ follows from (5.38) and (5.53). So the images of circumferences

$$x^2 + y^2 = (A + B)^2 \quad \text{and} \quad x^2 + y^2 = (a + b)^2$$

under the mapping (5.40) are the ellipses Γ_1 and Γ_2, respectively. Hence the function (5.40) maps conformally the ring

$$(a + b)^2 < x^2 + y^2 < (A + B)^2 \tag{5.54}$$

onto the domain D bounded by the ellipses Γ_1 and Γ_2.

This means that the domain D coincides with the ring (5.54) within conformal mapping accuracy. Hence, according to Theorem 5.1 and formula (5.19), we get

Theorem 5.3 *Let the cross-section of a cylindrical capacitor K be bounded by the confocal ellipses (5.36) and (5.37). Then the capacitance C of this capacitor is defined by*

$$C = 2\pi\varepsilon\varepsilon_0 \cdot \left(ln \frac{A+B}{a+b} \right)^{-1}. \tag{5.55}$$

3. Let Γ_1 and Γ_2 denote the ellipses (5.36) and (5.37). Consider the mapping

$$z = \zeta^{-1}. \tag{5.56}$$

Let $z = x + iy$ and $\zeta = \xi + i\eta$, then

$$\xi = \frac{x}{x^2 + y^2}, \quad \eta = -\frac{y}{x^2 + y^2}. \tag{5.57}$$

Substituting ξ and η from (5.57) into (5.36) and (5.37) we get

$$L_1 : \quad \frac{x^2}{A^2} + \frac{y^2}{B^2} = (x^2 + y^2)^2, \tag{5.58}$$

$$L_2 : \quad \frac{x^2}{a^2} + \frac{y^2}{b^2} = (x^2 + y^2)^2, \tag{5.59}$$

where $A > B,\ a > b,\ A > a,\ B > b$ and $A^2 - B^2 = a^2 - b^2$.

The function (5.56) maps conformally the complex plane without point O onto itself, the ellipses Γ_1 and Γ_2 onto the curves L_1 and L_2, respectively. Hence this function maps conformally the domain D_1 bounded by the ellipses Γ_1 and Γ_2 onto the domain G bounded by the ellipses L_1 and L_2.

Thus, using Theorems 5.1 and 5.3 we get

Theorem 5.4 *Let the cross-section of a capacitor K be bounded by the curves (5.58) and (5.59). Then the capacitance C of this capacitor is defined by*

$$C = 2\pi\varepsilon\varepsilon_0 \left(ln \frac{A+B}{a+b} \right)^{-1}. \tag{5.60}$$

4. Consider now the circumferences Γ_1 and Γ_2 :

$$\Gamma_1 : \quad (\xi + \frac{d}{2})^2 + \eta^2 = r^2, \tag{5.61}$$

$$\Gamma_2 : \quad (\xi - \frac{d}{2})^2 + \eta^2 = R^2, \tag{5.62}$$

where

$$d > 0, \quad 0 < r < R, \quad r > \frac{d}{2}, \quad R > r + d. \tag{5.63}$$

The circumference Γ_1 under the condition (5.63) is included in the circle

$$(\xi - \frac{d}{2})^2 + \eta^2 < R^2.$$

Consider the mapping

$$z = \frac{1}{2}(\zeta + \frac{\alpha^2}{\zeta}), \tag{5.64}$$

where $0 < \alpha \leq r - \frac{d}{2}$ and $z = x + iy$, $\zeta = \xi + i\eta$.

As we have pointed out above, the function (5.64) maps conformally the domain $\xi^2 + \eta^2 > \alpha^2$ onto the complement of segment $[-\alpha, \alpha]$. We are going to write the parametric equations of the images of circumferences Γ_1 and Γ_2 under the mapping (5.64). Denote these images by L_1 and L_2, respectively.

The parametric equations of Γ_1 and Γ_2 in complex form look as follows

$$\Gamma_1: \quad \zeta = -\frac{d}{2} + re^{i\varphi}, \quad 0 \leq \varphi \leq 2\pi, \tag{5.65}$$

$$\Gamma_2: \quad \zeta = \frac{d}{2} + Re^{i\varphi}, \quad 0 \leq \varphi \leq 2\pi. \tag{5.66}$$

Substituting the parametric equation (5.65) into (5.64) we get

$$z = \frac{1}{2}(-\frac{d}{2} + re^{i\varphi} + \frac{\alpha^2}{-\frac{d}{2} + re^{i\varphi}}), \quad 0 \leq \varphi \leq 2\pi. \tag{5.67}$$

Equalizing the real and imaginary parts in (5.67) we get a parametric equation of the curve L_1 :

$$L_1: \begin{cases} x = \frac{1}{2}(-\frac{d}{2} + r\cos\varphi + \frac{\alpha^2(r\cos\varphi - \frac{d}{2})}{r^2 + \frac{d^2}{4} - rd\cos\varphi}), \\ \\ y = \frac{1}{2}(r\sin\varphi - \frac{r\alpha^2\sin\varphi}{r^2 + \frac{d^2}{4} - rd\cos\varphi}), \end{cases} \tag{5.68}$$

where $0 \leq \varphi \leq 2\pi$.

In a similar way we get a parametric equation of the curve L_2 in the form

$$L_2: \begin{cases} x = \frac{1}{2}(R\frac{d}{2} + R\cos\varphi + \frac{\alpha^2(R\cos\varphi + \frac{d}{2})}{r^2 + \frac{d^2}{4} + rd\cos\varphi}), \\ \\ y = \frac{1}{2}(R\sin\varphi - \frac{\alpha^2 R\sin\varphi}{R^2 + \frac{d^2}{4} + Rd\cos\varphi}), \end{cases} \tag{5.69}$$

where $0 \leq \varphi \leq 2\pi$.

Let D_0 be the domain bounded by circumferences Γ_1 and let Γ_2 and G be the domain bounded by curves L_1 and L_2. It is clear that the function (5.64) maps conformally the domain D_0 onto the domain G. This means that D_0 and G are equal within conformal mapping accuracy. Let D be the domain bounded by circumferences (5.30) and (5.31). The function $z = \zeta - \frac{d}{2}$ maps conformally the domain D_0 onto the domain D. So the domains G and D are equal within conformal mapping accuracy.

Hence, applying Theorems 5.1 and 5.2 we get

Theorem 5.5 *Let the cross-section G of a cylindrical capacitor K be bounded by curves (5.68) and (5.69). Then its capacitance is defined by (5.32), where R_1, R_2 and x_0 are defined by (5.33) - (5.35).*

The domain G depends on four real variables r, R, d and α. In particular, when

$$R = A + B, \quad r = a + b, \quad d = 0, \quad \alpha = \sqrt{A^2 - B^2}, \quad A^2 - B^2 = a^2 - b^2$$

the domain G coincides with the domain, bounded by confocal ellipses (5.36) and (5.37). In the case where $d \neq 0$, the boundary of the domain G is different from ellipses.

5. Consider now the circumferences

$$(\xi - x_0)^2 + \eta^2 = R^2, \tag{5.70}$$

$$(\xi - x_0)^2 + \eta^2 = r^2, \tag{5.71}$$

where $0 < r < R < x_0$.

Consider the conformal mapping $z = \sqrt{\zeta}$, i.e.

$$\xi = x^2 - y^2, \quad \eta = 2xy.$$

Substituting ξ and η into (5.70) and (5.71) we obtain

$$(x^2 + y^2)^2 - 2x_0(x^2 - y^2) + x_0^2 - R^2 = 0, \quad x > 0, \tag{5.72}$$

$$(x^2 + y^2)^2 - 2x_0(x^2 - y^2) + x_0^2 - r^2 = 0, \quad x > 0. \tag{5.73}$$

Applying Theorem 5.1 and formula (5.19) we get

Theorem 5.6 *Let the cross-section of a cylindrical capacitor K be bounded by the curves (5.72) and (5.73) ($0 < r < R < x_0$). Then the capacitance of this capacitor is defined by*

$$C = 2\pi\varepsilon\varepsilon_0 \cdot \left(ln\frac{R}{r}\right)^{-1}.$$

6. Consider now the ring $r^2 < \xi^2 + \eta^2 < R^2$ and the mapping

$$z = \mu(\zeta + \alpha\zeta^n), \tag{5.74}$$

where $z = x + iy$, $\zeta = \xi + i\eta$ and

$$0 < r < R \leq 1, \quad \mu > 0, \quad 0 < \alpha < \frac{1}{n}.$$

The function (5.74) maps conformally this ring onto the biconnected domain D bounded by curves γ_1 and γ_2

$$\gamma_1: \ x = \mu(R\cos\varphi + \alpha R^n \cos n\varphi), \quad y = \mu(R\sin\varphi + \alpha R^n \sin\varphi), \tag{5.75}$$

$$\gamma_2: \ x = \mu(r\cos\varphi + \alpha r^n \cos n\varphi), \quad y = \mu(r\sin\varphi + \alpha r^n \sin\varphi), \tag{5.76}$$

where $0 \leq \varphi \leq 2\pi$.

From Theorem 5.1 and formula (5.19) we get

Theorem 5.7 *Let the cross-section of a cylindrical capacitor K be bounded by the curves (5.75) and (5.76) ($\mu > 0$, $0 < \alpha < \frac{1}{n}$, $0 < r < R \leq 1$). Then the capacitance of this capacitor is defined by*

$$C = 2\pi\varepsilon\varepsilon_0 \cdot \left(ln\frac{R}{r}\right)^{-1}.$$

7. Consider now the domain G_0 bounded by the confocal ellipses

$$\Gamma_1: \quad \frac{\xi^2}{A^2} + \frac{\eta^2}{B^2} = 1, \tag{5.77}$$

$$\Gamma_2: \quad \frac{\xi^2}{a^2} + \frac{\eta^2}{b^2} = 1, \tag{5.78}$$

where

$$0 < a < A < 1, \quad B > b, \quad A > B, \quad A^2 - B^2 = a^2 - b^2.$$

The parametric equations of ellipses Γ_1 and Γ_2 have the form

$$\Gamma_1: \quad \xi = A\cos\varphi, \quad \eta = B\sin\varphi, \tag{5.79}$$

$$\Gamma_2: \quad \xi = a\cos\varphi, \quad \eta = b\sin\varphi, \tag{5.80}$$

where $0 \leq \varphi \leq 2\pi$.

Let $z = x + iy$ and $\zeta = \xi + i\eta$. Denote by L_1 and L_2 the images of Γ_1 and Γ_2 under the mapping

$$z = (\zeta + \alpha\zeta^2)\mu \quad (0 < \alpha < \frac{1}{2}).$$

From (5.79) and (5.80) it follows that the parametric equations of these curves L_1 and L_2 have the form

$$\begin{cases} x = \mu(A\cos\varphi + \alpha A^2\cos^2\varphi - \alpha B^2\sin^2\varphi), \\ y = \mu(A\sin\varphi + 2\alpha AB\sin\varphi\cos\varphi), \end{cases} \tag{5.81}$$

and

$$\begin{cases} x = \mu(a\sin\varphi + \alpha a^2\cos^2\varphi - \alpha b^2\sin^2\varphi), \\ y = \mu(a\sin\varphi + 2\alpha ab\sin\varphi\cos\varphi), \end{cases} \tag{5.82}$$

where $0 \leq \varphi \leq 2\pi$.

The function $\zeta = (z + \alpha z^2)\mu$ maps conformally the domain G_0 onto the domain G bounded by curves L_1 and L_2.

From Theorems 5.1 and 5.4 we get

Theorem 5.8 *Let* $0 < a < A < 1$, $B > b$, $A > B$, $a > b$, $A^2 - B^2 = a^2 - b^2$, *and* $0 < \alpha < 1$, *and let the cross-section of a cylindrical capacitor be bounded by the curves (5.81) and (5.82). Then the capacitance of this capacitor is defined by (5.55).*

Consider now the following curves

$$\frac{(\xi - x_0)^2}{a^2} + \frac{\eta^2}{b^2} = [(\xi - x_0)^2 + \eta^2]^2, \tag{5.83}$$

$$\frac{(\xi - x_0)^2}{A^2} + \frac{\eta^2}{B^2} = [(\xi - x_0)^2 + \eta^2]^2, \tag{5.84}$$

where

$$A > a, \quad B > b, \quad a > b, \quad A > B, \quad A^2 - B^2 = a^2 - b^2, \quad x_0 > \frac{1}{a}. \tag{5.85}$$

In (5.83) and (5.84) changing the variable

$$\xi = x^2 - y^2, \quad \eta = 2xy, \quad (x \geq 0),$$

we get

$$\frac{(x^2 - y^2 - x_0^2)}{a^2} + \frac{4x^2y^2}{b^2} = [(x^2 - y^2 - x_0)^2 + 4x^2y^2]^2, \tag{5.86}$$

$$\frac{(x^2 - y^2 - x_0^2)}{A^2} + \frac{4x^2y^2}{B^2} = [(x^2 - y^2 - x_0)^2 + 4x^2y^2]^2. \tag{5.87}$$

Let D be the domain bounded by curves (5.58) and (5.59) and let G be the domain bounded by curves (5.86) and (5.87). It is clear that these two domains are equal within conformal mapping accuracy.

Hence, applying Theorems 5.1 and 5.4 we get

Theorem 5.9 *Let the condition (5.85) be satisfied and let the cross-section of a cylindrical capacitor K be bounded by curves (5.86) and (5.87). Then the capacitance of this capacitor is defined by*

$$C = 2\pi\varepsilon\varepsilon_0 \cdot \left(ln \frac{A+B}{a+b} \right)^{-1}.$$

5.4 Some Tests of Equality for Capacitances

In section 5.2 we have obtained a necessary and sufficient condition of equality of capacitances for cylindrical capacitors. Here we are mentioning some simple tests of equality of capacitances for such capacitors. These tests permit us to describe the shape of cylindrical capacitor with given capacitance.

1. Let the cross-section of a cylindrical capacitor K_0 be bounded by the curves Γ_1 and Γ_2:

$$\Gamma_1 : f_1(\xi, \eta) = 0, \tag{5.88}$$

$$\Gamma_2 : f_2(\xi, \eta) = 0. \tag{5.89}$$

Without loss of generality we assume that the curves (5.88) and (5.89) are in the half-plane $\xi > 0$. This can always be attained by switching the co-ordinate system.

Consider the conformal mapping

$$z = \sqrt[n]{\zeta}, \tag{5.90}$$

where $z = x + iy$, $\zeta = \xi + i\eta$; n is a positive integer. Here

$$\xi = P_n(x, y), \quad \eta = Q_n(x, y), \tag{5.91}$$

where $P_n(x, y) = Re(x + iy)^n$ and $Q_n(x, y) = Im(x + iy)^n$.

Substituting ξ and η from (5.91) into (5.88) and (5.89) we get

$$f_1(P_n(x, y), Q_n(x, y)) = 0, \tag{5.92}$$

$$f_2(P_n(x, y), Q_n(x, y)) = 0, \tag{5.93}$$

where

$$x > 0, \quad | \arctan \frac{y}{x} | \leq \frac{\pi}{n}.$$

From Theorem 5.1 we have

Theorem 5.10 *Let K_0 be a capacitor with cross-section bounded by the curves (5.88) and (5.89) and let K be a capacitor with cross-section bounded by the curves (5.92) and (5.93). Then the capacitances of these capacitors are equal.*

2. Let the cross-section D_0 of cylindrical capacitor K_0 be bounded by the smooth closed curves γ_1 and γ_2 :

$$\gamma_1 : z \equiv x + iy = \psi_1(\theta), \quad 0 \leq \theta \leq 2\pi, \tag{5.94}$$

$$\gamma_2 : z \equiv x + iy = \psi_2(\theta), \quad 0 \leq \theta \leq 2\pi, \tag{5.95}$$

where

$$\psi_j(\theta) = \alpha_j(\theta) + i\beta_j(\theta), \quad j = 1, 2. \tag{5.96}$$

Let $f_0(z)$ be an arbitrary analytic function in the domain D_0 with continuous derivative $f_0'(z)$ in closed domain $\bar{D}_0 = D_0 \cup \gamma_1 \cup \gamma_2$. Then

$$| f(z_1) - f(z_2) | \leq \alpha_0 \mid z_1 - z_2 \mid, \quad z_1 \in \bar{D}_0, \quad z_2 \in \bar{D}_0, \tag{5.97}$$

where α_0 is a positive constant, depending on $f(z)$ and D_0 and is independent of z_1 and z_2.

If $f(z)$ is of the form

$$f(z) = \sum_{k=0}^{n} [c_k z^k + d_k (z - z_0)^{-k}],$$

the choice of α_0 implying the inequality (5.97) is not difficult.

Consider the mapping

$$\zeta = (z + \alpha f(z))\mu, \tag{5.98}$$

where α and μ are complex constants satisfying

$$| \alpha | \alpha_0 < 1, \quad \mu \neq 0, \tag{5.99}$$

where α_0 is the constant from (5.97).

Denote by Γ_1 and Γ_2 the images of the curves γ_1 and γ_2 under the mapping (5.98). Substituting the values of z from (5.94) and (5.95) into (5.98), we get the parametric equations of these curves in complex form

$$\Gamma_1 : \zeta \equiv \xi + i\eta = \mu(\psi_1(\theta) + \alpha f(\psi_1(\theta))), \tag{5.100}$$

$$\Gamma_2 : \zeta \equiv \xi + i\eta = \mu(\psi_2(\theta) + \alpha f(\psi_2(\theta))), \tag{5.101}$$

or in co-ordinates

$$\Gamma_1 : \xi = Re\Phi_1(\theta), \quad \eta = Im\Phi_1(\theta), \tag{5.102}$$

$$\Gamma_2 : \xi = Re\Phi_2(\theta), \quad \eta = Im\Phi_2(\theta), \tag{5.103}$$

where

$$\Phi_j(\theta) = \mu(\psi_j(\theta) + \alpha f(\psi_j(\theta))), \quad j = 1, 2.$$

Let us prove that under the condition (5.99), the mapping (5.98) is one-to-one from γ_1 onto Γ_1.

Indeed, we have $z_1 = x_1 + iy_1 \in \gamma_1$, $z_2 = x_2 + iy_2 \in \gamma_1$ and $z_1 \neq z_2$. The images of ζ_1 and ζ_2 under the mapping (5.98) belong to the contour Γ_1 and

$$\zeta_1 = \mu(z_1 + \alpha f(z_1)), \tag{5.104}$$

$$\zeta_2 = \mu(z_2 + \alpha f(z_2)). \tag{5.105}$$

Let us show that $\zeta_1 \neq \zeta_2$. To this end we consider the difference $\zeta_1 - \zeta_2$:

$$\zeta_2 - \zeta_1 = \mu[z_1 - z_2 + \alpha(f(z_2) - f(z_1))]. \tag{5.106}$$

From (5.106) we have

$$| \zeta_2 - \zeta_1 | \geq | \mu | (| z_2 - z_1 | - | \alpha || f(z_1) - f(z_2) |).$$

From here, using the inequalities (5.97) and (5.99) we get

$$| \zeta_2 - \zeta_1 | \geq | \mu || z_2 - z_1 | (1 - | \alpha | \alpha_0) \neq 0.$$

Thus, (5.98) maps one-to-one γ_1 onto Γ_1. In a similar way we can prove that (5.98) maps one-to-one γ_2 onto Γ_2.

Let D be the domain bounded by the curves Γ_1 and Γ_2. Since the right-hand side of (5.98) is analytic function, which maps the boundary of D_0 onto the boundary of D_1, it is conformal mapping from D_0 onto D (see [14]). Hence the domains D and D_0 are equal within conformal mapping accuracy.

Thus, from Theorem 5.1 we get

Theorem 5.11 *Let the cross-section of a cylindrical capacitor K_0 be bounded by the curves (5.94) and (5.95) and let the cross-section of a cylindrical capacitor K be bounded by the curves (5.102) and (5.103). Then the capacitances of these two capacitors are equal.*

3. Consider now the mapping

$$\zeta = (\frac{1}{z - \zeta_0} + \alpha f(z))\mu, \tag{5.107}$$

where $f(z)$ is the analytic function from (5.98) and $\zeta_0 = \xi_0 + i\eta_0$ is a fixed point not belonging to the closed domain $\bar{D}_0$. Recall that D_0 is the cross-section of the capacitor K_0.

Let $f(z)$ satisfy the Lipschitz condition (5.97) and let α and μ be such that

$$\alpha\alpha_0 d_0^2 < 1, \quad \mu \neq 0, \tag{5.108}$$

where

$$d_0 = \max_{z \in \overline{D_0}} | z - \zeta_0 | .$$

Denote by L_1 and L_2 the images of the curves (5.94) and (5.95) under the mapping (5.107). The parametric equations of these curves are defined by

$$L_1 : \xi = Rew_1(\theta), \quad \eta = Imw_1(\theta), \tag{5.109}$$

$$L_2 : \xi = Rew_2(\theta), \quad \eta = Imw_2(\theta), \tag{5.110}$$

where

$$w_j(\theta) = \mu \left(\frac{1}{\psi_j(\theta) - \zeta_0} + \alpha f(\psi_j(\theta)) \right),$$
$$\psi_j(\theta) = \alpha_j(\theta) + i\beta_j(\theta).$$

Using the above arguments we can prove

Theorem 5.12 *Let the cross-section of a cylindrical capacitor K_0 be bounded by the contours (5.94), (5.95). Then the capacitance of any cylindrical capacitor with cross-section bounded by the contours (5.109) and (5.110) is equal to the capacitance of the capacitor K_0.*

Thus, if the capacitance of K_0 is fixed, then (changing $f(z)$, α, μ and ζ_0) we can obtain the capacitances of different capacitors.

Consider now some applications of Theorems 5.11 and 5.12.

Example 5.1. Let the domain D_0 be the ring

$$r^2 < \xi^2 + \eta^2 < R^2.$$

Consider the mapping

$$z = (\zeta + \frac{\alpha}{\zeta - \delta})\mu, \tag{5.111}$$

where $z = x + iy$, $\zeta = \xi + i\eta$ and δ, α and μ are real constants such that $0 < \delta < r < R$ and

$$| \alpha |< (r - \delta)^2, \quad \mu \neq 0.$$

Denote by L_1 and L_2 the images of the circumferences $\xi^2 + \eta^2 = R^2$ and $\xi^2 + \eta^2 = r^2$ under the mapping (5.111). The parametric equations of these curves are defined by

$$L_1 : z = x + iy = (Re^{i\varphi} + \frac{\alpha}{Re^{i\varphi} - \delta})\mu, \tag{5.112}$$

$$L_2 : z = x + iy = (re^{i\varphi} + \frac{\alpha}{re^{i\varphi} - \delta})\mu, \tag{5.113}$$

where $0 \leq \varphi \leq 2\pi$.

Applying Theorem 5.11 and formula (5.19) we get

Corollary 5.1 *Let the cross-section of a cylindrical capacitor K be bounded by the curves (5.112) and (5.113). Then the capacitance of this capacitor is defined by (5.19) and the curves (5.112) and (5.113) for $\delta = 0$ are confocal ellipses.*

Example 5.2. Let D_0 be biconnected domain bounded by the confocal ellipses γ_1 and γ_2 :

$$\gamma_1 : \frac{\xi^2}{a^2} + \frac{\eta^2}{b^2} = 1,$$
$$\gamma_2 : \frac{\xi^2}{A^2} + \frac{\eta^2}{B^2} = 1,$$

where $A > a$, $B > b$, $A > B$, $a > b$, and $A^2 - B^2 = a^2 - b^2$.

Consider the mapping

$$z = \mu(\zeta + \frac{\alpha}{\zeta}), \tag{5.114}$$

where α and μ are real constants satisfying $\mid \alpha \mid < b^2$, $\mu \neq 0$.

Denote by L_1 and L_2 the images of γ_1 and γ_2 under the mapping (5.114). The parametric equations of these curves can be written in the form

$$L_1 : z = x + iy = \mu \left(a\cos\varphi + ib\cos\varphi + \frac{\alpha}{a\cos\varphi + ib\cos\varphi} \right), \tag{5.115}$$

$$L_2 : z = x + iy = \mu \left(A\cos\varphi + iB\cos\varphi + \frac{\alpha}{A\cos\varphi + iB\cos\varphi} \right), \tag{5.116}$$

where $0 \leq \varphi \leq 2\pi$.

Applying Theorem 5.11 and formula (5.55) we get

Corollary 5.2 *Let the cross-section of a cylindrical capacitor K be bounded by the curves (5.115) and (5.116). Then the capacitance of K is defined by (5.55).*

Example 5.3. Let the domain D_0 be bounded by γ_1 and γ_2 :

$$\gamma_1 : \xi^2 + \eta^2 = R^2,$$
$$\gamma_2 : (\xi + d)^2 + \eta^2 = r^2,$$

where $d > 0$, and $r + d < R$.

Consider the mapping

$$z = (\frac{1}{\zeta + d} + \alpha e^{\zeta})\mu, \tag{5.117}$$

where μ and α are real constants satisfying

$$\mu \neq 0, \quad |\alpha| < e^{-R}(R+d)^{-1}.$$

Let L_1 and L_2 be the images of the circumferences γ_1 and γ_2 under the mapping (5.117):

$$L_1 : z = x + iy = \mu \left[\frac{1}{Re^{i\varphi} + d} + \alpha \exp(Re^{i\varphi}) \right], \tag{5.118}$$

$$L_2 : z = x + iy = \mu \left[\frac{1}{re^{i\varphi}} + \alpha \exp(re^{i\varphi} - d) \right], \tag{5.119}$$

where $0 \leq \varphi \leq 2\pi$.

Applying Theorem 5.12 and formula (5.32) we get

Corollary 5.3 *Let the cross-section of a cylindrical capacitor K be bounded by the curves (5.118) and (5.119). Then the capacitance of this capacitor is defined by (5.32).*

5.5 A Method of Definition of Capacitance for Spherical Capacitors

Let a spherical capacitor K be bounded by the spheres

$$x^2 + y^2 + z^2 = R^2, \tag{5.120}$$

$$(x+d)^2 + y^2 + z^2 = r^2, \tag{5.121}$$

where R, r and d are positive constants satisfying $r + d < R$.

In the case where $d = 0$ the capacitance of K is defined by (see [19])

$$C = 4\pi\varepsilon\varepsilon_0 \frac{Rr}{R-r}. \tag{5.122}$$

In [19] and [22] the capacitances of the spherical capacitors are expressed by the numerical series. The terms of these series are functions *sh*, *ch*, *Arcsh*, *Arcch*.

Here we propose another, simple formula for capacitance of spherical capacitors permitting us to write approximative formulae of an arbitrary precision and to evaluate the corresponding errors.

We introduce some notation. Let

$$x_0 = \frac{R^2 - r^2 + d^2 + \sqrt{(R^2 - r^2 + d^2)^2 - 4R^2d^2}}{2d}, \tag{5.123}$$

$$x_1 = \frac{x_0}{x_0^2 - R^2}, \tag{5.124}$$

$$x_2 = \frac{x_0 - d}{(x_0 - d)^2 - r^2}, \tag{5.125}$$

$$R_1 = \frac{R}{x_0^2 - R^2}, \tag{5.126}$$

$$R_2 = \frac{r}{(x_0 - d)^2 - r^2}, \tag{5.127}$$

$$q = R_2 R_1^{-1}. \tag{5.128}$$

Theorem 5.13 *Let a spherical capacitor K be bounded by the spheres (5.120) and (5.121). Then the capacitance of this capacitor is defined by*

$$C = 4\pi\varepsilon\varepsilon_0 R_2 \sum_{k=0}^{\infty} \frac{q^k}{x_1^2 - q^{2k} R_2^2}, \tag{5.129}$$

where the constants x_1, R_2 and q are defined by (5.124), (5.127) and (5.128).

Further, we are going to show that

$$q = \frac{r x_0}{R(x_0 - d)}, \quad x_0 > R, \quad x_1 > R_2 > 0, \quad 0 < q < 1.$$

Since the series (5.129) converges, as n-th approximation for capacitance C we can take the sum of the first n terms of this series, i.e.

$$C_n = 4\pi\varepsilon\varepsilon_0 R_2 \sum_{k=0}^{n-1} \frac{q^k}{x_1^2 - q^{2k} R_2^2}. \tag{5.130}$$

From (5.129) and (5.130) we have

$$| C - C_n | \leq 4\pi\varepsilon\varepsilon_0 R_2 \frac{q^n}{(1-q)(x_1^2 - R_2^2 q^{2n})}. \tag{5.131}$$

It follows from (5.131) that for $n \to \infty$ the difference $C - C_n$ tends to zero as decreasing geometrical progression.

To prove Theorem 5.13 we need some auxiliary propositions. To each point P we correspond a point Q, which is symmetric to P with respect to the sphere $x^2 + y^2 + z^2 = 1$. These two points are on a ray with origin coinciding with the origin of co-ordinate system and $|\overrightarrow{OP}||\overrightarrow{OQ}|=1$, where $|\overrightarrow{OP}|$ and $|\overrightarrow{OQ}|$ are the distances between the origin of co-ordinate system and the points P and Q, respectively.

The co-ordinates (x,y,z) and (ξ,η,ζ) of P and Q, respectively, under such mapping are related by

$$\xi = \frac{x}{x^2+y^2+z^2}, \quad \eta = \frac{y}{x^2+y^2+z^2}, \quad \zeta = \frac{z}{x^2+y^2+z^2}. \tag{5.132}$$

The inverse mapping is defined by

$$x = \frac{\xi}{\xi^2+\eta^2+\zeta^2}, \quad y = \frac{\eta}{\xi^2+\eta^2+\zeta^2}, \quad z = \frac{\zeta}{\xi^2+\eta^2+\zeta^2}. \tag{5.133}$$

Lemma 5.1 *Let $x_0 > R$. Then the image of the sphere*

$$(\xi - x_0)^2 + \eta^2 + \zeta^2 = R^2 \tag{5.134}$$

under the mapping (5.132) is the sphere

$$(x - x_1)^2 + y^2 + z^2 = R_1^2, \tag{5.135}$$

where x_1 and R_1 are defined by (5.124) and (5.126).

Proof. Substituting ξ, η, and ζ from (5.132) into equation (5.134) we get

$$(x - x_0\rho^2)^2 + y^2 + z^2 = R^2\rho^4, \tag{5.136}$$

where $\rho^2 = x^2+y^2+z^2$.

Opening the parentheses in (5.136) and dividing both sides by ρ^2 we get equation (5.135).

Lemma 5.1 is proved.

Let the constants d, r, x_0 satisfy the conditions

$$d > 0, \quad r > 0, \quad x_0 > d + r.$$

Then, in a similar way, we get that the image of the sphere

$$(\xi - x_0 + d)^2 + \eta^2 + \zeta^2 = r^2 \tag{5.137}$$

under the mapping (5.132) is the sphere

$$(x - x_2)^2 + y^2 + z^2 = R_2^2, \tag{5.138}$$

where x_2 and R_2 are defined by (5.125) and (5.127).

Let $x_0 > R > r > 0$, $d > 0$ and $r + d < R$. Then under the condition

$$\frac{x_0}{x_0^2 - R^2} = \frac{x_0 - d}{(x_0 - d)^2 - r^2} \tag{5.139}$$

the centers of spheres (5.135) and (5.138) coincide.

Consider equation (5.139) with respect to x_0. This equation coincides with (5.27) and has two real solutions. One of these solutions is greater than R and is defined by (5.123).

Thus, we have obtained the following

Lemma 5.2 *Let D be the domain, bounded by the spheres (5.134) and (5.137) and let x_0 be the constant from (5.123). Then the image of D under the mapping (5.132) is the spherical ring*

$$G : R_2^2 < (x - x_1)^2 + y^2 + z^2 < R_1^2, \tag{5.140}$$

where x_1, R_1 and R_2 are defined by (5.124), (5.126) and (5.127), respectively.

The formulae (5.125), (5.126) and (5.127) can be written as follows

$$x_2 = \frac{1}{2}\left[\frac{1}{x_0 - d - r} + \frac{1}{x_0 - d + r}\right], \tag{5.141}$$

$$R_1 = \frac{1}{2}\left[\frac{1}{x_0 - R} - \frac{1}{x_0 + R}\right], \tag{5.142}$$

$$R_2 = \frac{1}{2}\left[\frac{1}{x_0 - d - r} - \frac{1}{x_0 - d + r}\right]. \tag{5.143}$$

Since $x_0 > R$ and $r + d < R$, then from (5.128), (5.139), (5.141) – (5.143) it follows that

$$0 < R_2 < R_1, \quad x_1 = x_2 > R_2, \quad 0 < q = \frac{r x_0}{R(x_0 - d)} < \frac{r}{R - d} < 1. \tag{5.144}$$

Lemma 5.3 *Let $v(x, y, z)$ be a harmonic in the domain*

$$(x - x_1)^2 + y^2 + z^2 > R_1^2,$$

and continuously differentiable in the closed domain $(x - x_1)^2 + y^2 + z^2 \geq R_1^2$ function such that $v(x, y, z) \to 0$ as $\rho \to +\infty$, where $\rho = \sqrt{x^2 + y^2 + z^2}$. Then the function

$$u(x, y, z) = \frac{1}{\rho} v\left(\frac{x}{\rho^2}, \frac{y}{\rho^2}, \frac{z}{\rho^2}\right) \tag{5.145}$$

satisfy the relation

$$\frac{1}{4\pi} \int_{\sigma_1} \frac{\partial u(p)}{\partial N_p} ds_p = -v(0, 0, 0), \tag{5.146}$$

where σ_1 is the sphere (5.135), N_p is the outer normal to σ_1 at the point P of integration, ds_p is the area element of σ_1.

Proof. Let $v(x, y, z)$ satisfy the conditions of Lemma 5.3 and let $u(x, y, z)$ be defined by (5.145). Denote by D_1 and G_1 the domains

$$D_1 : (x - x_0)^2 + y^2 + z^2 > R_1^2, \tag{5.147}$$

$$G_1 : (\xi - x_1)^2 + \eta^2 + \zeta^2 > R_2^2. \tag{5.148}$$

According to Lemma 5.1 the domains D_1 and G_1 are symmetric with respect to the sphere $x^2 + y^2 + z^2 = 1$. Observe that (5.145) is the Kelvin transform of $v(x, y, z)$. Since $v(x, y, z)$ is harmonic in G_1 and tends to zero as $\rho \to +\infty$, then the Kelvin transform (5.145) is also harmonic in the symmetric domain D_1 (see [25]). From (5.145) it follows that $u(x, y, z) \to 0$ as $\rho \to +\infty$.

Consider the integral

$$J_\nu \equiv \frac{1}{4\pi} \int\int_{\Omega_\nu} \frac{\partial u(p)}{\partial N_p} ds_p, \tag{5.149}$$

where Ω_μ is the sphere

$$x^2 + y^2 + z^2 = \nu^2, \quad \nu > x_0 + R_1,$$

and N_p is the outer normal to Ω_ν at the point p of integration. Since $u(x, y, z)$ is harmonic in the domain D_1 and tends to zero as $\rho \to +\infty$, then according to Green's formula, we have

$$\frac{1}{4\pi} \int_{\sigma_1} \frac{\partial u(p)}{\partial N_p} ds_p = J_\nu. \tag{5.150}$$

Using (5.150) we have

$$\frac{\partial u(p)}{\partial N_p} = \frac{\partial u(p)}{\partial \rho} = -\frac{1}{\nu^2} v(0,0,0) + O(\frac{1}{\nu^3}), \quad for \quad p \in \Omega_\nu. \tag{5.151}$$

From this relation we obtain

$$-\frac{1}{4\pi} \int_{\Omega_\nu} \frac{\partial u(p)}{\partial N_p} ds_p = +v(0,0,0) - \frac{1}{4\pi} \int_{\Omega_\nu} O(\frac{1}{\nu^3}) ds_p. \tag{5.152}$$

Hence

$$\lim_{\nu \to +\infty} J_\nu = -v(0,0,0). \tag{5.153}$$

Passing to the limit in (5.150) as $\nu \to +\infty$, we get the equality (5.146).

Lemma 5.3 is proved.

Consider the spherical ring

$$R_2^2 < x^2 + y^2 + z^2 < R_1^2, \quad (R_1 > R_2). \tag{5.154}$$

Denote by Ω_1 and Ω_2 the boundaries of this ring:

$$\begin{aligned} \Omega_1 : x^2 + y^2 + z^2 &= R_1^2, \\ \Omega_2 : x^2 + y^2 + z^2 &= R_2^2. \end{aligned}$$

Consider the following Dirichlet problem

Problem A. Find a solution $u(x, y, z)$ of the Laplace equation

$$\frac{\partial^2 w}{\partial x^2} + \frac{\partial^2 w}{\partial y^2} + \frac{\partial^2 w}{\partial z^2} = 0 \tag{5.155}$$

belonging to the spherical ring (5.154) and satisfying the boundary conditions

$$w(x, y, z) = u_0(x, y, z), \quad (x, y, z) \in \Omega_1, \tag{5.156}$$

$$w(x, y, z) = 0, \quad (x, y, z) \in \Omega_2, \tag{5.157}$$

where $u_0(x, y, z)$ is a given function which is harmonic in $x^2 + y^2 + z^2 < R_1^2$ and is continuous in the closed sphere.

Lemma 5.4 *A solution $u(x, y, z)$ of the problem A is defined by*

$$w(x, y, z) = w_1(x, y, z) - w_2(x, y, z), \tag{5.158}$$

where

$$w_1(x, y, z) = \sum_{k=0}^{\infty} q^k u_0(q^{2k}x, q^{2k}y, q^{2k}z), \tag{5.159}$$

$$w_2(x, y, z) = \sum_{k=0}^{\infty} R_2 q^k \frac{1}{\rho} u_0(R_2^2 q^{2k} \frac{x}{\rho^2}, R_2^2 q^{2k} \frac{y}{\rho^2}, R_2^2 q^{2k} \frac{z}{\rho^2}), \tag{5.160}$$

$$q = R_2 R_1^{-1}, \quad \rho = \sqrt{x^2 + y^2 + z^2}.$$

Proof. Let $p(x, y, z)$ be a point belonging to the spherical ring (5.154). Then the points P_k and Q_k $(k = 0, 1, \ldots)$ with co-ordinates

$$(q^{2k}x, q^{2k}y, q^{2k}z) \quad \text{and} \quad (R_2^2 q^{2k} \frac{x}{\rho^2}, R_2^2 q^{2k} \frac{y}{\rho^2}, R_2^2 q^{2k} \frac{z}{\rho^2})$$

belong to the sphere $x^2 + y^2 + z^2 < R_1^2$. On the other hand, each term of the series (5.160) is Kelvin transform of a harmonic function $u_0(x, y, z)$. So the terms of the series (5.159) and (5.160) are harmonic in the domain (5.154). Since $0 < q < 1$, the series (5.159), (5.160) converge uniformly in the closed domain $R_2^2 \leq x^2 + y^2 + z^2 \leq R_1^2$. Thus, the sums of these series are harmonic functions in the domain (5.154). It is clear that $\rho = R_1$ for $(x, y, z) \in \Omega_1$ and $\rho = R_2$ for $(x, y, z) \in \Omega_2$.

Finally, it is easy to check that the harmonic function $w(x, y, z)$ defined by (5.158) satisfies the boundary conditions (5.156) and (5.157).

Lemma 5.4 is proved.

Proof of Theorem 5.13. Let D and G be the domains defined in Lemma 5.2 and let σ_1, σ_2 and σ_1^*, σ_2^* be the boundaries of these domains:

$$\sigma_1 : \quad (x - x_0)^2 + y^2 + z^2 = R^2, \tag{5.161}$$

$$\sigma_2 : \quad (x - x_0 + d)^2 + y^2 + z^2 = r^2, \tag{5.162}$$

$$\sigma_1^* : \quad (x - x_1)^2 + y^2 + z^2 = R_1^2, \tag{5.163}$$

$$\sigma_2^* : \quad (x - x_1)^2 + y^2 + z^2 = R_2^2. \tag{5.164}$$

According to Lemma 5.3 the domains D and G are symmetric with respect to the sphere $x^2 + y^2 + z^2 = 1$.

Let $u(x, y)$ be a harmonic function in the domain D satisfying the boundary conditions

$$u(x, y, z) = 1 \quad \text{for} \quad (x, y, z) \in \sigma_1, \tag{5.165}$$

$$u(x, y, z) = 0 \quad \text{for} \quad (x, y, z) \in \sigma_2. \tag{5.166}$$

Further, let the spherical capacitor K be bounded by the spheres σ_1 and σ_2. Then the capacitance C of this capacitor is defined by (cf. [4], p. 102)

$$C = \varepsilon \varepsilon_0 \int\limits_{\sigma_1} \frac{\partial u(p)}{\partial N_p} ds_p, \tag{5.167}$$

where N_p is the outer normal to σ_1 at the point p of integration.

It is clear that the capacitance of the capacitor bounded by the spheres (5.120) and (5.121) coincides with the capacitance of the above mentioned capacitor K, because these two capacitors are geometrically identical.

Consider the function

$$v(x, y, z) = \frac{1}{\rho} u\left(\frac{x}{\rho^2}, \frac{y}{\rho^2}, \frac{z}{\rho^2}\right), \tag{5.168}$$

where $\rho = \sqrt{x^2 + y^2 + z^2}$.

Since $u(x, y, z)$ is harmonic in the domain D, $v(x, y, z)$ is harmonic in the domain G. From conditions (5.165) and (5.166) we have

$$v(x, y, z) = \frac{1}{\sqrt{x^2 + y^2 + z^2}}, \quad (x, y, z) \in \sigma_1^*, \tag{5.169}$$

$$v(x, y, z) = 0, \quad (x, y, z) \in \sigma_2^*. \tag{5.170}$$

We set

$$w(x,y,z) = v(x+x_1,y,z). \tag{5.171}$$

By (5.169) – (5.171) the function $w(x,y,z)$ satisfies problem A at

$$u_0(x,y,z) = \frac{1}{\sqrt{(x+x_1)^2+y^2+z^2}}. \tag{5.172}$$

According to Lemma 5.4 a solution of problem A is defined by (5.158). Substituting $w(x,y,z)$ from (5.158) into (5.171) we get

$$v(x,y,z) = v_1(x,y,z) - v_2(x,y,z), \tag{5.173}$$

where

$$v_1(x,y,z) = w_1(x-x_1,y,z), \quad v_2(x,y,z) = w_2(x-x_1,y,z). \tag{5.174}$$

From (5.168) we have

$$u(x,y,z) = \frac{1}{\rho}v(\frac{x}{\rho^2},\frac{y}{\rho^2},\frac{z}{\rho^2}), \tag{5.175}$$

where $\rho = \sqrt{x^2+y^2+z^2}$.

Substituting $v(x,y,z)$ from (5.173) into (5.175), we get

$$u(x,y,z) = u_1(x,y,z) - u_2(x,y,z), \tag{5.176}$$

where

$$u_j(x,y,z) = \frac{1}{\rho}v_j(\frac{x}{\rho^2},\frac{y}{\rho^2},\frac{z}{\rho^2}), \quad j=1,2. \tag{5.177}$$

It follows from (5.159), (5.160) and (5.174) that $v_1(x,y,z)$ is harmonic in $(x-x_1)^2+y^2+z^2 < R_1^2$ and $v_2(x,y,z)$ is harmonic in $(x-x_1)^2+y^2+z^2 > R_1^2$ and $v_2(x,y,z) \to 0$ as $\rho \to +\infty$.

Thus, according to Green's formula, we have

$$\int_{\sigma_1} \frac{\partial u_1(p)}{\partial N_p} ds_p = 0. \tag{5.178}$$

From Lemma 5.3 we have

$$\frac{1}{4\pi}\int_{\sigma_1} \frac{\partial u_2(p)}{\partial N_p} ds_p = -v_2(0,0,0). \tag{5.179}$$

Substituting $u(x,y,z)$ from (5.176) into (5.167) and using (5.178) and (5.179) we get

$$C = 4\pi\varepsilon\varepsilon_0 v_2(0,0,0). \tag{5.180}$$

From (5.174) we have $v_2(0,0,0) = w_2(-x_1,0,0)$. Hence

$$C = 4\pi\varepsilon\varepsilon_0 w_2(-x_1,0,0). \tag{5.181}$$

Substituting $w_2(x,y,z)$ from (5.160) into (5.181) and using (5.172) we obtain (5.129).

Theorem 5.11 is proved.

5.6 Approximative Formulae for Capacitance

Let D_j be the cross-section of the capacitor K_j and let Γ_j and γ_j be the outer and inner boundaries of the domain D_j $(j = 1,\ldots,n)$ respectively. Further, let the domain D_j be in the plane Π_j. The origin of co-ordinate system in Π_j will always be taken in simple-connected domain, bounded by the inner boundary γ_j.

If we consider simultaneously several cylindrical capacitors K_j, $j = 1,2,\ldots,$ n we can, without loss of generality, assume that all the planes Π_j, $j = 1,\ldots,n$ and the coordinate systems coincide.

To obtain approximative formulae for capacitances we prove the following

Lemma 5.5 *Let D_1 and D_2 be the cross-sections of the capacitors K_1 and K_2 and let C_1 and C_2 be their capacitances. If $D_1 \subset D_2$, then $C_1 \geq C_2$.*

Proof. Let $D_1 \subset D_2$. Denote by $\Gamma_1 \cup \gamma_1$ and $\Gamma_2 \cup \gamma_2$ the boundaries of the domains D_1 and D_2, respectively. Let the curves Γ_1 and Γ_2 coincide.

Denote by $u_1(x,y)$ and $u_2(x,y)$ harmonic functions in D_1 and D_2 respectively, satisfying the conditions

$$u_1(x,y) = 1, \quad (x,y) \in \Gamma_1, \tag{5.182}$$

$$u_1(x,y) = 0, \quad (x,y) \in \gamma_1, \tag{5.183}$$

$$u_2(x,y) = 1, \quad (x,y) \in \Gamma_1, \tag{5.184}$$

$$u_2(x,y) = 0, \quad (x,y) \in \gamma_2. \tag{5.185}$$

According to the maximum principle a non-constant harmonic function attains its maximum and minimum on the boundary of the domain. Hence from (5.182) – (5.185) we have

$$0 < u_1(x,y) < 1, \quad (x,y) \in D_1, \tag{5.186}$$

$$0 < u_2(x,y) < 1, \quad (x,y) \in D_2. \tag{5.187}$$

From (5.182), (5.184) and (5.186) we have

$$u_2(x,y) - u_1(x,y) = 0, \quad (x,y) \in \Gamma_1, \tag{5.188}$$
$$u_2(x,y) - u_1(x,y) > 0, \quad (x,y) \in \gamma_1. \tag{5.189}$$

Hence, from the maximum principle of harmonic functions we get

$$u_2(x,y) > u_1(x,y), \quad (x,y) \in D_1. \tag{5.190}$$

Let $(\xi,\eta) \in \Gamma_1$ and $(x,y) \in D_1$. Then from (5.182), (5.184), (5.186), (5.187) and (5.190) we have

$$0 < u_2(\xi,\eta) - u_2(x,y) < u_1(\xi,\eta) - u_1(x,y).$$

Thus

$$0 \le \frac{\partial u_2(\xi,\eta)}{\partial N} \le \frac{\partial u_1(\xi,\eta)}{\partial N}, \quad (\xi,\eta) \in \Gamma_1, \tag{5.191}$$

where $\frac{\partial u_j(\xi,\eta)}{\partial N}$, $j = 1,2$ is the derivative of a function $u_j(\xi,\eta)$ along the outer normal N at the point $(\xi,\eta) \in \Gamma_1$.

The capacitances C_1 and C_2 of K_1 and K_2 are defined by

$$C_j = \varepsilon\varepsilon_0 \int\limits_{\Gamma_1} \frac{\partial u_j(\xi,\eta)}{\partial N} ds \quad (j = 1,2). \tag{5.192}$$

In (5.192) we have taken into account that the contours Γ_1 and Γ_2 coincide.

From (5.191) and (5.192), $C_1 \ge C_2$ immediately follows. In a similar way one can prove this if γ_1 and γ_2 coincide.

We are going to prove the assertion of Lemma 5.5 in general. Denote by K_3 the capacitor with the cross-section, bounded by the contours Γ_2 and γ_1. Let C_3 be the capacitance of K_3.

By the above arguments we have $C_1 \ge C_3$ and $C_3 \ge C_2$. Hence $C_1 \ge C_2$.

Lemma 5.5 is proved.

Now let D, D_1 and D_2 be the cross-sections of the capacitors K, K_1 and K_2, respectively, and let C, C_1 and C_2 be their capacitances. Further, let

$$D_1 \subset D \subset D_2. \tag{5.193}$$

According to Lemma 5.5, we have

$$C_2 \le C \le C_1. \tag{5.194}$$

From here we obtain

$$\left| C - \frac{C_2 + C_1}{2} \right| \le \frac{C_1 - C_2}{2}. \tag{5.195}$$

Hence if the capacitances of K_1 and K_2 are known and their difference is sufficiently small, then as the approximative value of the capacitance C of the capacitor K we can take the number

$$C \approx \frac{C_1 + C_2}{2}. \tag{5.196}$$

Let d_1 and d_2 be the absolute and relative errors of the approximative formula (5.196). Recall that these errors are defined by

$$d_1 = \left| C - \frac{C_1 + C_2}{2} \right|, \quad d_2 = \frac{d_1}{C}, \tag{5.197}$$

where C is the exact value of the capacitance of the capacitor K.

From (5.194) and (5.195) we have

$$d_1 \leq \frac{1}{2}(C_1 - C_2), \tag{5.198}$$

$$d_2 \leq \frac{1}{2}(\frac{C_1}{C_2} - 1). \tag{5.199}$$

Thus, we have obtained

Theorem 5.14 *Let D, D_1 and D_2 be the cross-sections of the cylindrical capacitors K, K_1 and K_2 and let C, C_1 and C_2 be their capacitances. If $D_1 \subset D \subset D_2$, then the approximative value of C is defined by*

$$C \approx \frac{C_1 + C_2}{2}, \tag{5.200}$$

where the absolute and relative errors satisfy the inequalities (5.198), (5.199).

Now we give some examples of applications of approximative formula (5.200).

Example 5.4. Let K_1 and K_2 be cylindrical capacitors with cross-sections D_1 and D_2 :

$$D_1 : \quad r_2^2 \leq x^2 + y^2 \leq r_1^2,$$
$$D_2 : \quad R_2^2 \leq x^2 + y^2 \leq R_1^2,$$

where $0 < R_2 \leq r_2 < r_1 \leq R_1$ and the differences $R_1 - r_1$ and $r_2 - R_2$ are sufficiently small.

According to (5.19), the capacitances C_1 and C_2 of K_1 and K_2 are defined by

$$C_1 = 2\pi\varepsilon\varepsilon_0 \cdot \left(ln\frac{R_1}{R_2} \right)^{-1} \quad \text{and} \quad C_2 = 2\pi\varepsilon\varepsilon_0 \cdot \left(ln\frac{r_1}{r_2} \right)^{-1}. \tag{5.201}$$

Let the cross-section D of a cylindrical capacitor K satisfy the relation

$$D_1 \subset D \subset D_2. \tag{5.202}$$

Then, according to Theorem 5.14 the approximative value of capacitance C of the capacitor K is defined by (5.200), where C_1 and C_2 are defined by (5.201) and the absolute and relative errors of this approximative value satisfy the estimates (5.198), (5.199).

Example 5.5. Let D_1 and D_2 be the cross-sections of the cylindrical capacitors K_1 and K_2. Let the domain D_1 be bounded by circumferences

$$(x+d)^2 + y^2 = \mu_1^2 \quad \text{and} \quad x^2 + y^2 = \rho_2^2, \tag{5.203}$$

and the domain D_2 by circumferences

$$(x+d)^2 + y^2 = \mu_2^2 \quad \text{and} \quad x^2 + y^2 = \rho_1^2, \tag{5.204}$$

where $\mu_2 \leq \mu_1 < \rho_2 \leq \rho_1$, $\mu_1 + d < \rho_2$ and the differences $\rho_1 - \rho_2$ and $\mu_1 - \mu_2$ are sufficiently small.

Let the cross-section D of the cylindrical capacitor K satisfy the relation (5.202). Then, according to Theorem 5.14 the approximative value C of the capacitor K is defined by (5.200), where C_1 and C_2 are the capacitances of K_1 and K_2 (cf. Theorem 5.2).

Example 5.6. Let D_1 and D_2 be the cross-sections of the capacitors K_1 and K_2. Let the domain D_1 be bounded by ellipses

$$\frac{x^2}{a_1^2} + \frac{y^2}{b_1^2} = 1 \quad \text{and} \quad \frac{x^2}{A_2^2} + \frac{y^2}{B_2^2} = 1, \tag{5.205}$$

and the domain D_2 be bounded by ellipses

$$\frac{x^2}{a_2^2} + \frac{y^2}{b_2^2} = 1 \quad \text{and} \quad \frac{x^2}{A_1^2} + \frac{y^2}{B_1^2} = 1, \tag{5.206}$$

where

$$A_1 \geq B_1, \quad A_2 \geq B_2, \quad a_1 \geq b_1, \quad a_2 \geq b_2,$$

$$A_1 \geq A_2 > a_1 \geq a_2, \quad B_1 \geq B_2 > b_1 \geq b_2, \tag{5.207}$$

$$A_1^2 - B_1^2 = a_2^2 - b_2^2, \quad A_2^2 - B_2^2 = a_1^2 - b_1^2. \tag{5.208}$$

According to (5.55), the capacitances C_1 and C_2 of K_1 and K_2 are defined by

$$C_1 = 2\pi\varepsilon\varepsilon_0 \cdot \left(ln \frac{A_2 + B_2}{a_1 + b_1} \right)^{-1}, \tag{5.209}$$

$$C_2 = 2\pi\varepsilon\varepsilon_0 \cdot \left(ln \frac{A_1 + B_1}{a_2 + b_2} \right)^{-1}. \tag{5.210}$$

Let the differences $A_1 - A_2$, $B_1 - B_2$, $a_1 - a_2$ and $b_1 - b_2$ be sufficiently small and let the cross-section of a cylindrical capacitor K satisfy the relations (5.202). Then the approximative value C of the capacitor K is defined by (5.200), where C_1 and C_2 are defined by (5.209), (5.210).

Example 5.7. Let the cross-section D of a capacitor K be bounded by the ellipses

$$\frac{x^2}{A^2} + \frac{y^2}{B^2} = 1 \quad \text{and} \quad \frac{x^2}{a^2} + \frac{y^2}{b^2} = 1,$$

where

$$A > B, \quad a > b, \quad A > a, \quad B > b.$$

Denote

$$\alpha = \sqrt{A^2 - B^2}, \quad \beta = \sqrt{a^2 - b^2}.$$

Further, let the difference $\alpha - \beta$ be sufficiently small with respect to A and a. In the case, where $\alpha = \beta$ the capacitance C of K is defined by (5.55). Using this formula for $\alpha = \beta$ we get an approximative formula for the general case.

Let $\beta < \alpha < a$. Consider the domain D_1 bounded by the ellipses

$$\frac{x^2}{A^2} + \frac{y^2}{B^2} = 1 \quad \text{and} \quad \frac{x^2}{b^2 + \alpha^2} + \frac{y^2}{b^2} = 1$$

and let the domain D_2 bounded by the ellipses

$$\frac{x^2}{A^2} + \frac{y^2}{B^2} = 1 \quad \text{and} \quad \frac{x^2}{a^2} + \frac{y^2}{a^2 - \alpha^2} = 1.$$

Let K_1 and K_2 be cylindrical capacitors with the cross-sections D_1 and D_2 and let C_1 and C_2 be their capacitances. Then according to (5.55) we have

$$C_1 = \frac{2\pi\varepsilon\varepsilon_0}{ln\frac{A+B}{\sqrt{b^2+\alpha^2}+b}}, \quad C_2 = \frac{2\pi\varepsilon\varepsilon_0}{ln\frac{A+B}{a+\sqrt{a^2-\alpha^2}}}.$$

These formulae can be written as follows

$$C_1 = \frac{2\pi\varepsilon\varepsilon_0}{ln\frac{A+B}{\sqrt{a^2+\gamma^2+b}}}, \quad C_2 = \frac{2\pi\varepsilon\varepsilon_0}{ln\frac{A+B}{a+\sqrt{b^2-\gamma^2}}}, \tag{5.211}$$

where $\gamma^2 = \alpha^2 - \beta^2$.

It is easy to see that $D_1 \subset D \subset D_2$. Therefore, by Theorem 5.14, we obtain the approximative formula (5.200) for the capacitance C of capacitor K, where C_1 and C_2 are defined by (5.211).

It is clear that if $\alpha < \beta$, then $D_2 \subset D \subset D_1$, and arguing as above, we again get (5.200).

5.7 Optimal Choice of Cylindrical Capacitor's Shape

Consider the following confocal ellipses Γ_1 and Γ_2,

$$\Gamma_1 : \quad \frac{x^2}{A^2} + \frac{y^2}{B^2} = 1, \tag{5.212}$$

$$\Gamma_2 : \quad \frac{x^2}{a^2} + \frac{y^2}{b^2} = 1, \tag{5.213}$$

where

$$A^2 - B^2 = a^2 - b^2, \tag{5.214}$$

$$A > B, \quad a \geq b, \quad A > a, \quad B > b. \tag{5.215}$$

Denote

$$\alpha = \sqrt{A^2 - B^2} = \sqrt{a^2 - b^2}.$$

Let K be a cylindrical capacitor with the cross-section bounded by the ellipses Γ_1 and Γ_2 and let C be the capacitance of K.

Consider the following two problems.

Problem A_1. For a given ellipse Γ_2 find a confocal ellipse Γ_1 such that

$$C = C_0, \tag{5.216}$$

where C_0 is a given positive number.

Problem A_2. For a given ellipse Γ_1 find a confocal ellipse Γ_2 such that

$$C = C_0. \tag{5.217}$$

Theorem 5.15 *The problem A_1 is uniquely solvable.*

Theorem 5.16 *The problem A_2 is solvable if and only if*

$$\frac{B}{A} \geq \frac{\gamma^2 - 1}{\gamma^2 + 1}, \tag{5.218}$$

where

$$\gamma = \exp\left(\frac{2\pi\varepsilon\varepsilon_0}{C_0}\right). \tag{5.219}$$

If the condition (5.218) is satisfied, then the problem A_2 is uniquely solvable.

Proof of Theorem 5.15. According to (5.55) we have

$$C = 2\pi\varepsilon\varepsilon_0 \cdot \left(ln \frac{A + B}{a + b}\right)^{-1}. \tag{5.220}$$

From here and (5.217) we obtain

$$A + B = \gamma(a + b), \tag{5.221}$$

where γ is defined by (5.219).

We rewrite the condition (5.214) as follows

$$(A - B)(A + B) = a^2 - b^2. \tag{5.222}$$

Substituting $A + B$ from (5.221) into (5.222) we get

$$A - B = \frac{a - b}{\gamma}. \tag{5.223}$$

From (5.221) and (5.223) we obtain

$$A = \frac{1}{2}\left[\gamma(a + b) + \frac{a - b}{\gamma}\right], \quad B = \frac{1}{2}\left[\gamma(a + b) - \frac{a - b}{\gamma}\right]. \tag{5.224}$$

Since $\gamma > 1$, it follows from (5.224) that $A > a$, $B > b$ and $A \geq B > 0$. So the axes of ellipse Γ_1 are defined by (5.224).

Theorem 5.15 is proved.

Proof of Theorem 5.16. From (5.212) we have

$$a^2 = b^2 + \alpha^2,$$

where $\alpha = \sqrt{A^2 - B^2}$. Hence, (5.220) can be written as

$$C = 2\pi\varepsilon\varepsilon_0 \cdot \left(ln\frac{A + B}{\sqrt{b^2 + \alpha^2} + b}\right)^{-1}. \tag{5.225}$$

Therefore

$$C \geq 2\pi\varepsilon\varepsilon_0 \cdot \left(ln\frac{A + B}{\alpha}\right)^{-1} = 4\pi\varepsilon\varepsilon_0 \cdot \left(ln\frac{A + B}{A - B}\right)^{-1}. \tag{5.226}$$

Substituting C from (5.217) into (5.226) we get

$$C_0 \geq 4\pi\varepsilon\varepsilon_0 \cdot \left(ln\frac{A + B}{A - B}\right)^{-1}. \tag{5.227}$$

The inequality (5.227) is equivalent to the inequality (5.218). Therefore, (5.218) is necessary for the solvability of problem A_2. Let this condition be satisfied. Then from (5.221) and (5.223) we have

$$a = \frac{1}{2}\left[\frac{A + B}{\gamma} + (A - B)\gamma\right] \quad \text{and} \quad b = \frac{1}{2}\left[\frac{A + B}{\gamma} - (A - B)\gamma\right]. \tag{5.228}$$

From (5.218) and (5.228) it follows that $a < A$, $b < B$ and $a > b > 0$. So under the condition (5.218) the axes of ellipse Γ_2 are defined by (5.228).

Theorem 5.16 is proved.

Now we have the following two curves L_1 and L_2 :

$$L_1: \quad \frac{x^2}{a^2} + \frac{y^2}{b^2} = (x^2 + y^2)^2, \tag{5.229}$$

$$L_2: \quad \frac{x^2}{A^2} + \frac{y^2}{B^2} = (x^2 + y^2)^2, \tag{5.230}$$

where

$$A \geq B, \quad a \geq b, \quad A > a, \quad B > b, \quad A^2 - B^2 = a^2 - b^2. \tag{5.231}$$

The curves L_1 and L_2 are symmetric to ellipses (5.212) and (5.213) with respect to circumference $x^2 + y^2 = 1$.

Let K be the cylindrical capacitor with the cross-section bounded by the curves L_1 and L_2 and let C be its capacitance.

Problem B_1. For a given contour L_2 find a contour L_1 such that $C = C_0$ where C_0 is a given positive number.

Problem B_2. For a given contour L_1 find a contour L_2 such that $C = C_0$.

Theorem 5.17 *The problem B_1 is solvable if and only if the condition (5.218) is satisfied, in this case the problem B_1 is uniquely solvable.*

Theorem 5.18 *The problem B_2 is uniquely solvable.*

Proof of Theorems 5.17 and 5.18 is similar to that of Theorems 5.15 and 5.16.

Theorems 5.15 – 5.18 show that for a given capacitance and for a given boundary surface one can find another boundary surface.

Let K be a cylindrical capacitor with the cross-section bounded by the confocal ellipses (5.212) and (5.213) and let C be the capacitance of this capacitor.

Consider the following problems

Problem A_3. Find confocal ellipses Γ_1 and Γ_2 of form (5.212) and (5.213) satisfying the conditions

$$C = C_0 \tag{5.232}$$

$$A - a = d_0, \quad \frac{b}{a} = \mu_0, \tag{5.233}$$

where C_0, d_0 and μ_0 are given positive constants, $\mu_0 \leq 1$. Notice that if the conditions (5.214) and (5.215) are satisfied, then $A - a$ is the shortest distance between the ellipses Γ_1 and Γ_2.

Theorem 5.19 *The problem A_3 is uniquely solvable.*

Proof. Using (5.232) we obtain (5.224), where γ is defined by (5.219). Resolving the system (5.224), (5.233) with respect to A, B, a and b we get

$$a = \frac{2d_0\gamma}{(\gamma-1)^2 + \mu_0(\gamma^2-1)}, \tag{5.234}$$

$$b = \frac{2d_0\mu_0\gamma}{(\gamma-1)^2 + \mu_0(\gamma^2-1)}, \tag{5.235}$$

$$A = \frac{d_0[\gamma^2(1+\mu_0) + 1 - \mu_0]}{(\gamma-1)^2 + \mu_0(\gamma^2-1)}, \tag{5.236}$$

$$B = \frac{d_0[\gamma^2(1+\mu_0) + \mu_0 - 1]}{(\gamma-1)^2 + \mu_0(\gamma^2-1)}. \tag{5.237}$$

Since $\gamma > 1$, $0 < \mu_0 \leq 1$, then from (5.234) – (5.237) it follows that the constants a, b, A and B are positive and satisfy the inequalities (5.215). Thus, the axes of ellipses (5.212) and (5.213) are defined by (5.234) – (5.237).

Theorem 5.19 is proved.

Let K be a cylindrical capacitor with the cross-section bounded by the curves (5.229) and (5.230) and let C be its capacitance.

Consider the following

Problem B_3. Find contours of the form (5.229) and (5.230) satisfying the conditions

$$C = C_0 \tag{5.238}$$

$$\frac{1}{a} - \frac{1}{A} = d_0, \quad \frac{b}{a} = \mu_0, \tag{5.239}$$

where C_0, d_0 and μ_0 are positive constants and $\mu_0 \leq 1$.

Notice that under the condition (5.231) the difference $\frac{1}{a} - \frac{1}{A}$ is the shortest distance between the curves (5.229) and (5.230).

Theorem 5.20 *The problem B_3 is uniquely solvable.*

Proof. Using (5.60) and (5.238) we get the relations (5.224). Resolving the system (5.224), (5.239) with respect to A, B, a and b we get

$$a = \frac{(1-\gamma)^2 + \mu_0(\gamma^2-1)}{d_0[(1+\mu_0)\gamma^2 + 1 - \mu_0]}, \tag{5.240}$$

$$b = \frac{\mu_0[(1-\gamma)^2 + \mu_0(\gamma^2-1)]}{d_0[(1+\mu_0)\gamma^2 + 1 - \mu_0]}, \tag{5.241}$$

$$A = \frac{(1-\gamma)^2 + \mu_0(\gamma^2-1)}{2\gamma d_0}, \tag{5.242}$$

$$B = \frac{[(1-\gamma)^2 + \mu_0(\gamma^2-1)][\gamma^2 - 1 + \mu_0(\gamma^2+1)]}{2\gamma d_0[(1+\mu_0)\gamma^2 + 1 - \mu_0]}. \tag{5.243}$$

Since $\gamma > 1$, $0 < \mu_0 \leq 1$, then it follows from (5.240) – (5.243) that the constants A, B, a and b are positive and satisfy the inequalities (5.231).

Theorem 5.20 is proved.

Consider the following

Problem A_4. Find confocal ellipses of the form (5.212) and (5.213) satisfying the conditions

$$C = C_0 \tag{5.244}$$

$$A - a = d_0, \quad \frac{B}{A} = \mu_0, \tag{5.245}$$

where C_0, d_0 and μ_0 are given positive constants, $\mu_0 \leq 1$ and C is the capacity of a cylindrical capacitor K with the cross-section bounded by the ellipses (5.212) and (5.213).

Assuming solvability of this problem, we are going to find the values of μ_0.

Theorem 5.21 *The problem A_4 is solvable if and only if*

$$\mu_0 \geq \frac{\gamma^2 - 1}{\gamma^2 + 1}, \tag{5.246}$$

where γ is defined by (5.219). Under the condition (5.246) the problem A_4 is uniquely solvable.

Proof. Let the problem A_4 have a solution and let A, B, a and b be axes of the contours Γ_1 and Γ_2. Then by (5.227) and by the second condition in (5.245) we have

$$C_0 \geq 4\pi\varepsilon\varepsilon_0 \cdot \left(ln \frac{1 + \mu_0}{1 - \mu_0} \right)^{-1}. \tag{5.247}$$

The inequality (5.247) is equivalent to (5.246). Hence the condition (5.246) is necessary for the solvability of the problem A_4.

Let this condition be satisfied. Then resolving the system of equations (5.228) and (5.245) with respect to A, B, a and b we get

$$a = \frac{\nu_1 d_0}{1 - \nu_1}, \quad b = \frac{\nu_2 d_0}{1 - \nu_1}, \quad A = \frac{d_0}{1 - \nu_1}, \quad B = \frac{\mu_0 d_0}{1 - \nu_1}, \tag{5.248}$$

where

$$\nu_1 = \frac{1 + \mu_0 + \gamma^2(1 - \mu_0)}{2\gamma}, \quad \nu_2 = \frac{1 + \mu_0 - \gamma^2(1 - \mu_0)}{2\gamma}. \tag{5.249}$$

It follows from (5.246) that $0 < \nu_1 < 1$, $0 < \nu_2 \leq \nu_1$, $0 < b \leq a$, $0 < B \leq A$, $a < A$, and $b < B$. So under the condition (5.246) the problem A_4 is uniquely solvable and the solution is defined by (5.248).

Theorem 5.21 is proved.

Problem B_4. Find contours of the form (5.229) and (5.230) satisfying the conditions

$$C = C_0 \tag{5.250}$$

$$\frac{1}{a} - \frac{1}{A} = d_0, \quad \frac{B}{A} = \mu_0, \tag{5.251}$$

where C_0, d_0 and μ_0 are positive constants and $\mu_0 \leq 1$. Here C is the capacitance of a cylindrical capacitor K with the cross-section bounded by the curves (5.229) and (5.230).

Theorem 5.22 *The problem B_4 is solvable if and only if μ_0 satisfies the inequality (5.246), in this case the problem B_4 is uniquely solvable.*

Theorem 5.22 is proved as Theorem 5.21. The constants A, B, a and b determining the curves (5.229), (5.230) are defined by

$$a = \frac{1-\nu_1}{d_0}, \quad b = \frac{\nu_2(1-\nu_1)}{\nu_1 d_0}, \quad A = \frac{1-\nu_1}{\nu_1 d_0}, \quad B = \frac{\mu_0(1-\nu_1)}{\nu_1 d_0}, \tag{5.252}$$

where ν_1 and ν_2 are the same as in (5.248).

Problem A_5. Let K be a cylindrical capacitor with the cross-section bounded by the ellipses (5.212) and (5.213), and let C be its capacitance. Find a capacitor with the least area of base satisfying the following conditions

$$C = C_0, \tag{5.253}$$

$$A - a \geq d, \quad \mu_2 \leq \frac{b}{a} \leq \mu_1, \tag{5.254}$$

$$A^2 - B^2 = a^2 - b^2, \tag{5.255}$$

$$A \geq B, \quad a \geq b, \quad A > a, \quad B > b, \tag{5.256}$$

where C_0, d, μ_1 and μ_2 are positive constants and $\mu_2 \leq \mu_1 \leq 1$.

Theorem 5.23 *The problem A_5 is uniquely solvable and the solution coincides with the solution of the problem A_3 at $\mu_0 = \mu_1$, $d_0 = d$.*

Proof. Denote

$$A - a = d_0, \quad \frac{b}{a} = \mu_0. \tag{5.257}$$

It follows from (5.253) that

$$d_0 \geq d, \quad \mu_2 \leq \mu_0 \leq \mu_1. \tag{5.258}$$

Substituting C from (5.252) into (5.220) we get

$$C_0 = 2\pi\varepsilon\varepsilon_0 \cdot \left(ln\frac{A+B}{a+b} \right)^{-1}. \tag{5.259}$$

We resolve the system of equations (5.254), (5.256) and (5.258) with respect to A, B, a and b. A solution of this system is defined by (5.234) – (5.237) (cf. Problem A_3), where γ is defined by (5.219).

The area of the base S of a capacitor K is defined by

$$S = \pi AB. \tag{5.260}$$

Substituting A and B from (5.236) and (5.237) into (5.259) we get

$$S = \pi d_0^2 f(\mu_0), \tag{5.261}$$

where

$$f(x) = \frac{\gamma^4(1+x)^2 - (1-x)^2}{[\gamma - 1 + x(\gamma+1)]^2(\gamma-1)^2}. \tag{5.262}$$

The derivative of $f(x)$ is defined by

$$f'(x) = -\frac{4\gamma[(\gamma^3+1)x + \gamma^3 - 1]}{(\gamma-1)^2[(\gamma+1)x + \gamma - 1]^3}. \tag{5.263}$$

Since $\gamma > 1$, it follows from (5.262) that $f'(x) < 0$ for $0 \le x \le 1$.

So the function $f(x)$ decreases in the segment $\mu_2 \le x \le \mu_1$ and reaches its smallest value at the point $x = \mu_1$. Since $d_0 = d$ and $\mu_0 = \mu_1$, we have

$$A - a = d, \quad \frac{b}{a} = \mu_1.$$

Thus, a solution of the problem A_5 coincides with the solution of the problem A_3 at $d_0 = d$ and $\mu_0 = \mu_1$. Hence the axes of the ellipses (5.212) and (5.213) are defined by (5.234) – (5.237), with d and μ_1 instead of d_0 and μ_0.

Theorem 5.23 is proved.

Problem A_6. Let K be a cylindrical capacitor with the cross-section bounded by the ellipses (5.212) and (5.213). Let the axes A, B, a and b of these ellipses and the capacitance C of a capacitor K satisfy the conditions (5.252) – (5.255). Find a capacitor with the least mass per unit length.

Theorem 5.24 *The problem A_6 is uniquely solvable and the solution coincides with the solution of the problem A_3 at $\mu_0 = \mu_1$ and $d_0 = d$.*

Proof. The mass m per unit length of a capacitor K is defined by

$$m = \pi\rho(AB - ab), \tag{5.264}$$

where ρ is the density of mass of K. Using (5.234) – (5.237) we get

$$m = \pi\rho d_0^2(f(\mu_0) - f_0(\mu_0)), \tag{5.265}$$

where $f(x)$ is defined by (5.261) and

$$f_0(x) = \frac{4\gamma^2 x}{(\gamma-1)^2[(\gamma+1)x+\gamma-1]^2}. \tag{5.266}$$

From (5.262) and (5.265) we have

$$f'(x) - f_0'(x) = -\frac{4\gamma(\gamma+1)[x(\gamma-1)+\gamma+1]}{(\gamma-1)[x(\gamma+1)+\gamma-1]^3}. \tag{5.267}$$

Since $\gamma > 1$, (5.266) implies

$$f'(x) - f_0'(x) < 0 \quad \text{for} \quad \mu_2 \le x \le \mu_1.$$

Hence the function $f(x) - f_0(x)$ in the segment $\mu_2 \le x \le \mu_1$ reaches its minimum at the point $x = \mu_1$. Thus, from (5.264) and (5.257) it follows that the mass of K reaches its minimum at $d_0 = d$, $\mu_0 = \mu_1$, i.e.

$$A - a = d, \quad \frac{a}{b} = \mu_1.$$

Thus, a solution of the problem A_6 coincides with that of problem A_3 at $\mu_0 = \mu_1$, $d_0 = d$ and the axes of the ellipses (5.212) and (5.213) are defined by (5.234) – (5.237), where $d_0 = d$ and $\mu_0 = \mu_1$.

Theorem 5.25 is proved.

Now we formulate a similar problem of the optimal choice of a capacitor K with the cross section, bounded by the curves (5.229) and (5.230), where A, B, a and b satisfy the condition (5.231). To this end we prove the following

Lemma 5.6 *The area S_1 bounded by the curve (5.229) is defined by*

$$S_1 = \frac{\pi}{2}(\frac{1}{a^2} + \frac{1}{b^2}). \tag{5.268}$$

Proof. Let (ρ, ψ) be the polar co-ordinates of a point belonging to the curve (5.229). The cartesian (x, y) and polar (ρ, ψ) co-ordinates are related by

$$x = \rho\cos\psi, \quad y = \rho\sin\psi. \tag{5.269}$$

Substituting x and y from (5.269) into (5.229) we get

$$L_1: \quad \rho = \sqrt{\frac{\cos^2\psi}{a^2} + \frac{\sin^2\psi}{b^2}}, \quad 0 \le \psi \le 2\pi. \tag{5.270}$$

Observe that (5.270) is the equation of the curve L_1 in polar co-ordinates. It is known that the area S_1 is defined by

$$S_1 = \frac{1}{2}\int_0^{2\pi} \rho^2 d\psi. \tag{5.271}$$

Substituting ρ from (5.269) into (5.270) and integrating, we obtain (5.267).

Lemma 5.6 is proved.

Note that by formula (5.268), the mass m per unit length of a capacitor K is defined by

$$m = \frac{\pi\gamma}{2}\left(\frac{1}{a^2} + \frac{1}{b^2} - \frac{1}{A^2} + \frac{1}{B^2}\right), \tag{5.272}$$

where γ is the mass density of a capacitor K.

Consider the following two problems.

Problem B_5. Let the cross-section of a capacitor K be bounded by the curves of the form (5.229) and (5.230). Find the parameters of capacitor K with the least area of base and satisfying the conditions

$$C = C_0, \quad A^2 - B^2 = a^2 - b^2, \tag{5.273}$$

$$\frac{1}{a} - \frac{1}{A} \ge d, \quad \mu_2 \le \frac{a}{b} \le \mu_1, \tag{5.274}$$

$$A \ge a, \quad B \ge b, \quad a > b, \quad A > B, \tag{5.275}$$

where d, μ_1 and μ_2 are given positive constants and $\mu_2 < \mu_1 \le 1$.

Problem B_6. Let the cross-section of a capacitor K be bounded by the curves of the form (5.229) and (5.230). Find the parameters of capacitor S with the smallest mass per unit length, satisfying the conditions (5.272) – (5.274).

Theorem 5.25 *The problems B_5 and B_6 are uniquely solvable and their solutions coincide with the solution of the problem B_3 at $\mu_0 = \mu_1$ and $d_0 = d$.*

Theorem 5.25 can be proved in the same way as Theorems 5.23 and 5.24. Instead of (5.234) – (5.237), (5.259) and (5.263) we need to use the formulae (5.240) – (5.243), (5.267) and (5.271).

Chapter 6

Efficient Methods of Solution of Boundary Value Problems for Improperly Elliptic Equations

6.1 Introduction

Let D be a simple connected bounded domain in the plane (OXY) with boundary Γ.

We consider the elliptic differential equation

$$Lu \equiv \sum_{k=0}^{n} A_k \frac{\partial^n u}{\partial x^k \partial y^{n-k}} = 0, \quad (x, y) \in D, \tag{6.1}$$

where A_k are complex constants.

The equation

$$A_0 \lambda^n + A_1 \lambda^{n-1} + \cdots + A_n = 0 \tag{6.2}$$

is called the *characteristic* equation of (6.1).

If equation (6.2) has no real roots and $A_0 \neq 0$, then (6.1) is said to be *elliptic.*

Equation (6.1) is called *properly elliptic,* if it is elliptic and the numbers of roots of the characteristic equation (6.2) belonging to $Im\lambda > 0$ and $Im\lambda < 0$

are equal, otherwise the elliptic equation (6.1) is called *improperly elliptic.*

It is clear that a properly elliptic equation is of even order.

Equation (6.1) is called *strongly elliptic,* if there exists a complex number β such that the inequality

$$Re\left[\beta\sum_{k=0}^{n}A_kx^ky^{n-k}\right]>0 \quad \text{for} \quad x^2+y^2\neq 0. \tag{6.3}$$

is valid. It is clear that the property of proper ellipticity of equation (6.1) follows from the strong ellipticity of (6.1). The opposite is not always the case. If n is even and the coefficients of the elliptic equation (6.1) are real, then this equation is strongly elliptic.

Let n be an even number. As the boundary conditions for Dirichlet problem of elliptic equation (6.1) we take

$$\frac{\partial^k u(x,y)}{\partial N^k}=f_k(x,y), \quad (x,y)\in\Gamma, \quad k=0,1,\ldots,\frac{n}{2}-1, \tag{6.4}$$

where $f_k(x,y)$ $(k=0,\ldots,\frac{n}{2}-1)$ are given complex valued functions and N is the exterior normal to the boundary Γ at the point $(x,y)\in\Gamma$.

In this chapter we will assume that the boundary Γ of the domain D and all the given functions defined on this boundary are infinitely differentiable, and the solutions are sought in the class of functions which are infinitely differentiable in the closed domain $D\bigcup\Gamma$.

The following two propositions take place (cf., e.g., [16], [26], [27]).

Proposition 6.1 *If equation (6.1) is properly elliptic, then the problem (6.1), (6.4) is Fredholmian.*

Proposition 6.2 *If equation (6.1) is strongly elliptic, then the problem (6.1), (6.4) is uniquely solvable.*

The assertions of Propositions 1 and 2 remain valid for bounded multi-connected domains.

It is known (see [16]), that the homogeneous problem (6.1), (6.4) (with $f_k\equiv 0$, $k=0,\ldots,\frac{n}{2}-1$) can possess an infinite number of linearly independent solutions, provided that the elliptic equation (6.1) is not properly elliptic.

The purpose of this chapter is to formulate and efficiently solve Dirichlet type correct boundary value problems for improperly elliptic equations of the form (6.1).

6.2 Some Auxiliary Propositions

The results of this section are of auxiliary character and will be used in the following two sections to study the boundary problems for improperly elliptic equations.

Consider a polynomial

$$u(x,y)=\sum_{k=0}^{n}\sum_{j=0}^{k}a_{jk}z^k\bar{z}^j, \quad z=x+iy, \quad \bar{z}=x-iy, \tag{6.5}$$

where a_{jk} are complex constants.

Lemma 6.1 *The polynomial (6.5) can be represented in the form*

$$u(x,y)=\sum_{k=0}^{n}\sum_{j=0}^{n-k}c_{kj}\bar{z}^j(z\bar{z}-1)^k, \tag{6.6}$$

where c_{kj} are some complex constants.

Proof. It is easy to see that the polynomial (6.5) can be represented in the form

$$u(x,y)=\sum_{k=0}^{n}\sum_{j=0}^{n-k}b_{jk}(z\bar{z})^k\bar{z}^j, \tag{6.7}$$

where b_{jk} are some complex constants.

According to binomial formula, we have

$$(z\bar{z})^k=(z\bar{z}-1+1)^k=\sum_{l=0}^{k}\frac{k!}{l!(k-l)!}(z\bar{z}-1)^l. \tag{6.8}$$

Substituting $(z\bar{z})^k$ from (6.8) into (6.7) we obtain (6.5).

Lemma 6.1 is proved.

Now let us consider the function

$$u(x,y)=\varphi_0(z)+\bar{z}\varphi_1(z)+\cdots+\bar{z}^{n-1}\varphi_{n-1}(z), \tag{6.9}$$

where $\varphi_0(z),\ldots,\varphi_{n-1}(z)$ are analytic functions in the circle $|z|<1$ ($|z|=\sqrt{x^2+y^2}$).

Lemma 6.2 *The function (6.9) can be represented in the form*

$$u(x,y)=\sum_{k=0}^{n-1}(z\bar{z}-1)^k\psi_k(z)+\bar{z}\sum_{k=0}^{n-2}\sum_{j=0}^{n-2-k}c_{kj}(z\bar{z}-1)^k\bar{z}^j, \tag{6.10}$$

where $\psi_k(z)$ $(k=0,\ldots,n-1)$ are some analytic functions in the circle $|z|<1$ and c_{jk} are some complex constants.

Proof. Let us represent the analytic functions $\psi_k(z)$ $(k = 1, \ldots, n-1)$ in the form

$$\varphi_k(z) = \sum_{j=0}^{k-1} a_{kj} z^j + z^k \alpha_k(z), \quad k = 1, \ldots, n, \tag{6.11}$$

where a_{jk} are complex constants and $\alpha_k(z)$ $(k = 1, \ldots, n)$ are analytic functions in the circle $| z |< 1$.

Substituting $\varphi_k(z)$ from (6.11) into (6.9) we get

$$u(x, y) = \varphi_0(z) + \sum_{k=1}^{n-1} (z\bar{z})^k \alpha_k(z) + \bar{z} \sum_{k=1}^{n-1} \sum_{j=0}^{k-1} a_{kj} \bar{z}^{k-1} z^j. \tag{6.12}$$

It is clear that

$$\sum_{k=1}^{n-1} \sum_{j=0}^{k-1} a_{kj} \bar{z}^{k-1} z^j \equiv \sum_{k=0}^{n-2} \sum_{j=0}^{k} a_{k+1,j} \bar{z}^k z^j. \tag{6.13}$$

According to Lemma 6.1 the right-hand side of (6.13) can be written in the form

$$\sum_{k=0}^{n-2} \sum_{j=0}^{k} a_{k+1,j} \bar{z}^k z^j = \sum_{k=0}^{n-2} \sum_{j=0}^{n-2-k} c_{kj} (z\bar{z} - 1)^k \bar{z}^j. \tag{6.14}$$

Using (6.13) and (6.14) the representation (6.12) can be written in the form

$$u(x, y) = \varphi_0(z) + \sum_{k=1}^{n-1} (z\bar{z})^k \alpha_k(z) + \bar{z} \sum_{k=1}^{n-2} \sum_{j=0}^{n-2-k} c_{kj} (z\bar{z} - 1)^k \bar{z}^j. \tag{6.15}$$

Substituting $(z\overline{z})^k$ from (6.8) into (6.15) we obtain (6.10), where $\psi_0(z), \ldots, \psi_{n-1}(z)$ are linearly expressed by $\varphi_0(z), \alpha_1(z), \ldots, \alpha_{n-1}(z)$.

Lemma 6.2 is proved.

Now let us consider a function of the form:

$$\begin{aligned} u(x, y) \quad = \quad & \varphi_0(z) + \bar{z}\varphi_1(z) + \cdots + \bar{z}^{n-1}\varphi_{n-1}(z) \\ & + \overline{\phi_0(z)} + z\overline{\phi_1(z)} + \cdots + z^{m-1}\overline{\phi_{m-1}(z)}, \end{aligned} \tag{6.16}$$

where $\varphi_0(z), \ldots, \varphi_{n-1}(z)$ and $\phi_0(z), \ldots, \phi_{n-1}(z)$ are analytic functions in $| z |< 1$ and $\overline{\phi_k(z)}$ is the complex conjugate to $\phi_k(z)$, $n \geq m \geq 1$.

Lemma 6.3 *If either $n = m$ or $n = m + 1$, then the function (6.16) can be represented as follows*

$$u(x, y) = \sum_{k=0}^{n-1} \alpha_k(z)(z\bar{z} - 1)^k + \sum_{k=0}^{m-1} \overline{\beta_k(z)}(z\bar{z} - 1)^k. \tag{6.17}$$

If $n \geq m+2$, then (6.16) can be represented in the form

$$u(x,y) = \sum_{k=0}^{n-1} \alpha_k(z)(z\overline{z}-1)^k + \sum_{k=0}^{m-1} \overline{\beta_k(z)}(z\overline{z}-1)^k + (z\overline{z}-1)^m \sum_{k=1}^{p}\sum_{j=0}^{k-1} c_{kj} z^j \overline{z}^k, \tag{6.18}$$

where $\alpha_0(z), \ldots, \alpha_{n-1}(z), \beta_0(z), \ldots, \beta_{m-1}(z)$ are some analytic in $|z|<1$ functions, c_{kj} are some complex constants and $p = n-m-1$.

Proof. Let $n \geq m+2$. Denote

$$v_1(x,y) = \varphi_0(z) + \varphi_1(z)\overline{z} + \ldots + \varphi_{n-1}(z)\overline{z}^{n-1},$$
$$v_2(x,y) = \phi_0(z) + \phi_1(z)\overline{z} + \ldots + \phi_{m-1}(z)\overline{z}^{m-1}.$$

The function (6.16) can be written as follows

$$u(x,y) = v_1(x,y) + \overline{v_2(x,y)}. \tag{6.19}$$

According to Lemma 6.2 the functions $v_1(x,y)$ and $v_2(x,y)$ can be represented in the form

$$v_1(x,y) = \sum_{k=0}^{n-1} \psi_{1k}(z)(z\overline{z}-1)^k + \overline{z}\sum_{k=0}^{n-2}\sum_{j=0}^{n-2-k} c_{1kj}(z\overline{z}-1)^k\overline{z}^j, \tag{6.20}$$

$$v_2(x,y) = \sum_{k=0}^{m-1} \psi_{2k}(z)(z\overline{z}-1)^k + \overline{z}\sum_{k=0}^{m-2}\sum_{j=0}^{m-2-k} c_{2kj}(z\overline{z}-1)^k\overline{z}^j, \tag{6.21}$$

where $\psi_{1k}(z)$ and $\psi_{2l}(z)$ $(k = 0,\ldots,n-1;\ l = 0,\ldots,m-1)$ are some analytic functions in the circle $|\,z\,|< 1$, and c_{1kj} and c_{2kj} are some complex constants. If $m = 1$, then $v_2(x,y) = \psi_0(z)$.

Substituting $v_1(x,y)$ and $v_2(x,y)$ from (6.20) and (6.21) into (6.19) we obtain the representation (6.18), where

$$\alpha_k(z) = \psi_{1k}(z) + z\sum_{k=0}^{m-2-k} \overline{c_{2kj}} z^j, \quad (k = 0,\ldots,m-1),$$

$$\alpha_k(z) = \psi_{1k}, \quad k = m,\ldots,n-1,$$

$$\beta_k(z) = \psi_{2k}(z) + z\sum_{j=0}^{n-2-k} \overline{c_{1kj}} z^j, \quad (k = 0,\ldots,m-1),$$

$$\sum_{k=m}^{n-2}\sum_{j=0}^{n-2-k} c_{1kj}(z\overline{z}-1)^{k-m}\overline{z}^{j+1} \equiv \sum_{k=1}^{p}\sum_{j=0}^{k-1} c_{kj} z^j \overline{z}^k, \quad p = n-m-1.$$

For $n = m$ (or $n = m + 1$) the assertion of Lemma 6.3 can be proved in a similar way.

Lemma 6.3 is proved.

Note 6.1. Without loss of generality, in (6.17) and (6.18) we can take

$$\beta_k(0) = 0, \quad k = 0, 1, \ldots, m - 1. \tag{6.22}$$

In (6.17) and (6.18) functions $\alpha_k(z)$ and $\beta_k(z)$ for $k = 0, \ldots, m - 1$ can be replaced by $\alpha_k(z) + \overline{\beta_k(0)}$ and $\beta_k(z) - \beta_k(0)$.

Consider the equation

$$\sum (A_{jk} z_k + B_{jk} \overline{z}_k) = C_j, \quad j = 1, 2, \ldots, n, \tag{6.23}$$

where A_{jk}, B_{jk} and C_j are given complex constants and $z_k = x_k + iy_k$ ($k = 1, \ldots, n$) is a solution to be found.

Lemma 6.4 *If for any right-hand side $C_1, \ldots, C_n$ the system (6.23) is solvable, then it is uniquely solvable.*

Proof is obvious.

6.3 Riemann–Hilbert Type Problem for a Class of Improperly Elliptic Equations

Let $\lambda = i$ be n-multiple root of the characteristic equation (6.2), then equation (6.1) can be written as follows

$$\frac{\partial^n u}{\partial \overline{z}^n} = 0, \quad (x, y) \in D, \tag{6.24}$$

where

$$\frac{\partial}{\partial \overline{z}} \equiv \frac{1}{2}\left(\frac{\partial}{\partial x} + i\frac{\partial}{\partial y}\right) \quad \text{and} \quad \frac{\partial}{\partial z} \equiv \frac{1}{2}\left(\frac{\partial}{\partial x} - i\frac{\partial}{\partial y}\right).$$

Let D be the unit circle $x^2 + y^2 < 1$, and let Γ be the circumference $x^2 + y^2 = 1$. The Riemann–Hilbert boundary conditions for equation (6.24) are taken in the form

$$Re\frac{\partial^k u}{\partial N^k} = f_k(x, y), \quad (x, y) \in \Gamma, \quad k = 0, \ldots, n - 1, \tag{6.25}$$

where N is the exterior normal to the boundary Γ at the point $(x, y) \in \Gamma$ and $f_0(x, y), \ldots, f_{n-1}(x, y)$ are given infinitely differentiable real-valued functions on Γ.

The problem (6.24), (6.25) for $f_k \equiv 0,\ k = 0, 1, \ldots, n-1$ is called *homogeneous* problem.

Let

$$F_k(z) = \frac{1}{2\pi i} \int\limits_{|\zeta|=1} f_k(\zeta) \frac{\zeta + z}{\zeta - z} \cdot \frac{d\zeta}{\zeta}, \tag{6.26}$$

$$a_{jkl} = \frac{k!}{l!(k-l)!} \omega_j^{(k-l)}(1), \quad l \le k,$$

where $\omega_j(x) = (x^2-1)^j$, $\zeta = \xi + i\eta$, $f_k(\zeta) = f_k(\xi, \eta)$ and $\omega_j^{(k)}(1)$ is the k-th derivative of a function $\omega_j(x)$ at the point $x = 1$. Since $\omega_j(x)$ is a real-valued function, the constants a_{kjl} are real.

The following theorems take place (see [28]):

Theorem 6.1 *The non-homogeneous problem (6.24), (6.25) is always solvable and the corresponding homogeneous problem has exactly n^2 linearly independent solutions.*

Theorem 6.2 *A particular solution $u_0(x, y)$ of the non-homogeneous problem (6.24), (6.25) is defined by*

$$u_0(x, y) = \sum_{j=0}^{n-1} (z\overline{z} - 1)^j \varphi_j(z), \tag{6.27}$$

where the functions $\varphi_0(z), \varphi_1(z), \ldots, \varphi_{n-1}(z)$ are analytic in the circle $\mid z \mid < 1$ and are defined by the recurrent formula

$$\varphi_0(z) = F_0(x), \tag{6.28}$$

$$k! 2^k \varphi_k(z) + \sum_{j=0}^{k-1} \sum_{l=0}^{k} a_{kjl} z^l \varphi_j^{(l)}(z) = F_k(z), \quad k = 1, \ldots, n-1. \tag{6.29}$$

Theorem 6.3 *The general solution of the homogeneous problem (6.24), (6.25) is defined by*

$$\begin{aligned} u(x, y) &= \sum_{k=1}^{n-1} \sum_{l=0}^{k-1} \left[a_{kl}(z^k \overline{z}^l - \overline{z}^k z^l) + + i b_{kl}(z^k \overline{z}^l + \overline{z}^k z^l) \right] \\ &+ \sum_{k=0}^{n-1} i c_k z^k \overline{z}^k, \end{aligned} \tag{6.30}$$

where a_{kl}, b_{kl} and c_k are arbitrary real constants.

In this chapter linear dependence or independence of the solutions of boundary problems is considered over the field of real numbers. We impose this assumption, because the set of solutions of the considered homogeneous problems is a linear space over the field of real numbers (this is not the case for the field of complex numbers).

Theorem 6.1 follows from Theorems 6.2 and 6.3. Hence we are going to prove the last two theorems.

Proof of Theorem 6.2. A particular solution of the non-homogeneous problem (6.24), (6.25) is searched in the form (6.27), where $\varphi_0(z), ..., \varphi_{n-1}(z)$ are analytic in the circle $| z |< 1$ and infinitely differentiable in $| z |\leq 1$ functions to be found.

Substituting $u(x, y)$ from (6.27) into the boundary conditions (6.25) we get

$$Re\omega(z) = f_k(z), \quad k = 0, \ldots, n-1, \quad | z |= 1, \tag{6.31}$$

where

$$\omega_0(z) = \varphi_0(z), \tag{6.32}$$

$$\omega_k(z) = 2^k k! \left(\varphi_k(z) + \sum_{j=0}^{k-1} \sum_{l=0}^{k} a_{kjl} z^l \varphi_j^{(l)}(z) \right), \quad j = 1, ..., n-1. \tag{6.33}$$

Since the functions $\omega_0(z), \ldots, \omega_{n-1}(z)$ are analytic in the circle $| z |< 1$, they are defined from the conditions (6.31) by Schwartz formula (see [14])

$$\omega_k(z) = F_k(z) + ia_k, \quad k = 0, \ldots, n-1, \tag{6.34}$$

where the function $F_k(z)$ is defined by (6.26), and $a_0, \ldots, a_{n-1}$ are arbitrary real constants. In particular, in (6.34) we take $a_k = 0$ $(k = 0, \ldots, n-1)$. To determine the analytic functions $\varphi_0(z), \ldots, \varphi_{n-1}(z)$, from (6.32) – (6.34) we obtain the recurrent formulae (6.28) and (6.29).

Theorem 6.2 is proved.

Proof of Theorem 6.3. Let $u(x, y)$ be a solution of the homogeneous problem (6.24), (6.25) (with $f_k \equiv 0, k = 0, \ldots, n-1$) and let

$$v(x, y) = u(x, y) + \overline{u(x, y)}. \tag{6.35}$$

Then

$$\frac{\partial^n \overline{u(x, y)}}{\partial \overline{z}^n} = 0 \quad (x, y) \in D, \tag{6.36}$$

$$\frac{\partial^{2n} v(x, y)}{\partial z^n \partial \overline{z}^n} = 0 \quad (x, y) \in D. \tag{6.37}$$

Equation (6.37) can be written in the form of n-homogeneous equation, $\Delta^n v = 0$, where Δ is the Laplacian,

$$\Delta \equiv \frac{\partial^2}{\partial x^2} + \frac{\partial^2}{\partial y^2}.$$

The conditions (6.25) with $f_k = 0$ $(k = 0, \ldots, n-1)$ can be written in the form

$$\frac{\partial^k v(x,y)}{\partial N^k} = 0, \quad k = 0, \ldots, n-1. \tag{6.38}$$

Equation (6.37) is strongly elliptic, hence the Dirichlet homogeneous problem (6.37), (6.38) has only zero solution (see [26]). Therefore

$$u(x,y) + \overline{u(x,y)} \equiv 0, \quad (x,y) \in D. \tag{6.39}$$

The general solution of equation (6.24) is represented in the form (see [16])

$$u(x,y) = \sum_{k=0}^{n-1} \varphi(z)\overline{z}^k, \tag{6.40}$$

where $\varphi_0(z), \ldots, \varphi_{n-1}(z)$ are arbitrary analytic functions in the domain D.

Substituting $u(x,y)$ from (6.40) into (6.39) we get

$$\sum_{k=0}^{n-1} (\varphi_k(z)\overline{z}^k + \overline{\varphi_k(z)} z^k) \equiv 0, \quad (x,y) \in D. \tag{6.41}$$

Let us expand $\varphi_k(z)$ into Taylor series

$$\varphi_k(z) = \sum_{j=0}^{\infty} c_{kj} z^j, \quad | z |< 1. \tag{6.42}$$

Substituting $\varphi_k(z)$ from (6.42) into (6.41) we obtain

$$\sum_{k=0}^{n-1} \sum_{j=0}^{\infty} [c_{kj} z^j z^k + \overline{c}_{kj} \overline{z}^j z^k], \quad | z |< 1. \tag{6.43}$$

Equating the coefficients of linearly independent functions $z^m \overline{z}^l$ to zero, we obtain

$$c_{kj} = 0, \quad k = 0, \ldots, n-1, \quad j = n, n+1, \ldots, \tag{6.44}$$

$$c_{kj} + \overline{c}_{jk} = 0, \quad k = 0, \ldots, n-1, \quad j = 0, \ldots, n-1. \tag{6.45}$$

Let

$$c_{kj} = a_{kj} + ib_{kj}, \tag{6.46}$$

where a_{kj} and b_{kj} are real constants.

From (6.45) we obtain

$$c_{kk} = ic_k, \quad k = 0, \ldots, n-1, \tag{6.47}$$

$$a_{kj} = -a_{jk}, \quad b_{kj} = b_{jk}, \; j = 0, \ldots, k-1; \; k = 1, \ldots, n-1, \tag{6.48}$$

where $c_0, \ldots, c_{n-1}$ are real constants.

Substituting $\varphi_k(z)$ from (6.42) into (6.40) and using (6.44) - (6.48), we obtain (6.30).

Theorem 6.3 is proved.

It is easy to see that the set of functions

$$iz^\rho \overline{z}^\rho, \quad z^k \overline{z}^l - \overline{z}^k z^l, \quad i(z^k \overline{z}^l + \overline{z}^k z^l);$$
$$(\rho = 0, \ldots, n-1; \quad l = 0, 1, \ldots, k-1; \quad k = 1, \ldots, n-1).$$

is linearly independent and their number is equal to n^2. Hence, from (6.30) it follows that the homogeneous problem (6.24), (6.25) has exactly n^2 linearly independent solutions.

Theorems 6.1 and 6.2 are proved.

Now we indicate additional conditions under which the problem (6.24), (6.25) is uniquely solvable. For unique solvability of the above problem we take the additional conditions

$$\frac{\partial^{k+l} u(0,0)}{\partial z^k \partial \overline{z}^l} = d_{kl}, \quad l = 0, \ldots, k-1; \quad k = 1, \ldots, n-1, \tag{6.49}$$

$$Im \frac{\partial^{2k} u(0,0)}{\partial z^k \partial \overline{z}^k} = d_k, \quad k = 0, 1, \ldots, n-1, \tag{6.50}$$

where d_{kl} are complex and d_k are real constants.

It follows from Theorems 6.2 and 6.3 that the problem (6.24), (6.25) with the additional conditions (6.49), (6.50) is uniquely solvable. This result remains valid if in (6.49), (6.50) the point (0,0) is replaced by any fixed point (x_0, y_0) from $D \bigcup \Gamma$.

Note 6.2. Let D be an arbitrary bounded simple connected or multi-connected domain with a smooth boundary. Then, in a similar way, one can prove that the assertion of Theorem 6.3 remains valid in these domains.

6.4 Dirichlet Type Problem for Improperly Elliptic Equations

Let D be the circle $x^2+y^2<1$, and Γ its boundary. Let us consider the elliptic equation (6.1), where the roots of the characteristic equation (6.2) are i and $-i$ with multiplicities p and q $(p+q=n)$, respectively.

We express equation (6.1) as follows

$$\frac{\partial^n u}{\partial \overline{z}^p \partial z^q}=0, \quad (x,y)\in D. \tag{6.51}$$

Without loss of generality we assume that $p\geq q$. If $p\neq q$, then equation (6.51) is improperly elliptic.

The boundary conditions for equation (6.51) are taken in the form

$$\frac{\partial^k u(x,y)}{\partial N^k}=f_k(x,y), \quad (x,y)\in D, \quad k=0,\ldots,q-1, \tag{6.52}$$

$$Re\frac{\partial^k u(x,y)}{\partial N^k}=f_k(x,y), \quad (x,y)\in \Gamma, \quad k=q,\ldots,p-1, \tag{6.53}$$

where N is the exterior normal to the boundary Γ at the point (x,y), $f_k(x,y)$, $(k=0,\ldots,p-1)$ are given functions on Γ and $f_q(x,y),\ldots,f_{p-1}(x,y)$ are real-valued.

If n is even and $p=q$, then the conditions (6.53) are absent and the problem (6.51), (6.52) is a Dirichlet problem for p-harmonic equation. If $p=n$ and $q=0$, then the conditions (6.52) are absent and the problem (6.51), (6.52) is a Riemann–Hilbert problem for equation (6.24). The problem (6.51) - (6.53) with $f_k=0$ $(k=0,\ldots,p-1)$ is said to be homogeneous.

The following theorems take place (see [28]).

Theorem 6.4 *The non-homogeneous problem (6.51) - (6.53) is always solvable and the corresponding homogeneous problem has exactly $(p-q)^2$ linearly independent solutions.*

Theorem 6.5 *The general solution of the homogeneous problem (6.51) - (6.53) is defined by*

$$\begin{aligned} u(x,y) &= (z\overline{z}-1)^q\sum_{k=0}^{m}\sum_{l=0}^{k-1}[a_{kl}(z^k\overline{z}^l-\overline{z}^k z^l) \\ &+ ib_{kl}(z^k\overline{z}^l+\overline{z}^k z^l)]+(z\overline{z}-1)^q\sum_{k=0}^{m}ic_k z^k\overline{z}^k, \end{aligned} \tag{6.54}$$

where $m=p-q-1$, and a_{kl}, b_{kl} and c_k are arbitrary real constants.

Proof of Theorems 6.4 and 6.5. A particular solution of the problem (6.51) – (6.53) is sought in the form

$$u(x,y) = \sum_{k=0}^{p-1} \varphi_{1k}(z)(z\overline{z}-1)^k + \sum_{k=0}^{q-1} \overline{\varphi_{2k}(z)}(z\overline{z}-1)^k, \tag{6.55}$$

where $\varphi_{1k}(z)$ and $\varphi_{2l}(z)$ $(k = 0,\ldots,p-1;\ l = 0,\ldots,q-1)$ are analytic functions to be found and

$$\varphi_{2l}(0) = 0, \quad k = 0,1,\ldots,q-1. \tag{6.56}$$

Substituting $u(x,y)$ from (6.55) into the boundary conditions (6.52) and (6.53), we obtain

$$\omega_{1k}(z) + \overline{\omega_{2k}(z)} = f_k(z), \quad |\, z \,| = 1, \quad k = 0,\ldots,q-1, \tag{6.57}$$

$$Re\omega_{1k}(z) = f_k(z), \quad |\, z \,| = 1, \quad k = q,\ldots,p-1, \tag{6.58}$$

where

$$\omega_{10}(z) = \varphi_{10}(z), \quad \omega_{20}(z) = \varphi_{20}(z), \tag{6.59}$$

$$\omega_{\rho k}(z) = k!2^k\varphi_{\rho k}(z) + \sum_{j=0}^{k-1}\sum_{l=0}^{k} a_{kjl}z^l\varphi_{\rho j}^{(l)}(z), \quad \rho = 1,2, \tag{6.60}$$

$$k = 1,\ldots,q-1,$$

$$\omega_{1k}(z) = k!2^k\varphi_{1k}(z) + \sum_{j=0}^{k-1}\sum_{l=0}^{k} a_{kjl}z^l\varphi_{1j}^{(l)}(z) + \sum_{j=0}^{q-1}\sum_{l=0}^{k} a_{kjl}z^l\varphi_{2j}^{(l)}(z), \tag{6.61}$$

$$k = q,\ldots,p-1,$$

$f_k(z) = f_k(x,y)$ $(k = 0,\ldots,p-1)$ and the constants a_{kjl} are defined by (6.26). It is clear that $\omega_{1k}(z)$ and $\omega_{2l}(z)$ $(k = 0,\ldots,p-1;\ l = 0,\ldots,q-1)$ are analytic functions in D. From (6.56) we have

$$\omega_{2k}(0) = 0, \quad k = 0,\ldots,q-1.$$

From the boundary condition (6.57), the analytic functions $\omega_{1k}(z)$ and $\omega_{2k}(z)$ with $k = 0,\ldots,q-1$ are defined by (see [29]):

$$\omega_{1k}(z) = \frac{1}{2\pi i}\int\limits_{|\zeta|=1} \frac{f_k(\zeta)d\zeta}{\zeta - z}, \quad \omega_{2k}(z) = \frac{z}{2\pi i}\int\limits_{|\zeta|=1} \frac{\overline{f_k(\zeta)}d\zeta}{\zeta(\zeta - z)}, \tag{6.62}$$

$$k = 0,1,\ldots,q-1, \quad \zeta = \xi + i\eta, \quad z = x + iy, \quad |\, z \,| < 1.$$

From the boundary conditions (6.58) we have (see [14]),

$$\omega_{1k}(z) = F_k(z) + ic_k, \quad k = q, \dots, p-1, \quad |\, z \,| < 1, \tag{6.63}$$

where the functions $F_k(z)$ $(k = q, \dots, p-1)$ are defined by (6.26), and c_k are arbitrary real constants.

In particular, in (6.63) we can take $c_k = 0$ $(k = q, \dots, p-1)$. Thus, in (6.59) – (6.61) the analytic functions $\omega_{1k}(z)$ and $\omega_{2k}(z)$ can be considered to be known. It is easy to see that the system of equations (6.59) – (6.63) is uniquely solvable with respect to $\varphi_{1k}(z)$ and $\varphi_{2l}(z)$ $(k = 0, \dots, p-1;\ l = 0, \dots, q-1)$.

Substituting this solution into (6.55) we obtain a particular solution of the problem (6.51) – (6.53).

Now we prove that the number of linearly independent solutions of the corresponding homogeneous problem is equal to $(p-q)^2$.

In the circle $|\, z \,| < 1$ the general solution of equation (6.51) is defined by (cf. [11] and [16]),

$$u(x, y) = \sum_{k=0}^{p-1} \varphi_k(z)\bar{z}^k + \sum_{k=0}^{q-1} \overline{\psi_k(z) z^k}, \tag{6.64}$$

where $\varphi_k(z)$ and $\psi_k(z)$ are arbitrary analytic in the circle $|\, z \,| < 1$ functions.

According to Lemma 6.3 and Note 6.1 any function of the form (6.64) can be represented as

$$\begin{aligned} u(x, y) &= \sum_{k=0}^{p-1} \varphi_{1k}(z)(z\bar{z} - 1)^k + \sum_{k=0}^{q-1} \overline{\varphi_{2k}(z)}(z\bar{z} - 1)^k \\ &+ (1 + z\bar{z})^q \sum_{k=1}^{p-q-1} \sum_{j=0}^{k-1} c_{kj} z^j \bar{z}^k, \end{aligned} \tag{6.65}$$

where $\varphi_{1k}(z)$ and $\varphi_{2l}(z)$ $(k = 0, \dots, p-1;\ l = 0, \dots, q-1)$ are analytic functions in the circle $|\, z \,| < 1$,

$$\varphi_{2l}(0) = 0, \quad l = 0, \dots, q-1,$$

and c_{kj} are complex constants.

Substituting $u(x, y)$ from (6.65) into the boundary conditions (6.52) and (6.53) (at $f_k = 0$ $(k = 0, \dots, p-1)$) to determine the analytic functions $\varphi_{1k}(z)$ and $\varphi_{2l}(z)$ $(k = 0, \dots, q-1;\ l = 0, \dots, q-1)$ we obtain the same boundary problem as (6.57) and (6.58) with the only difference that the right-hand sides of (6.57) and (6.58) are linear functions on Rec_{jk} and Imc_{jk} $(j = 0, \dots, k-1;$ $k = 1, \dots, p-q-1)$. Substituting a solution of this boundary problem into (6.65), we see that the general solution of the homogeneous problem (6.51) –

(6.53) contains $(p-q)^2$ real constants. From this it follows that the number of linearly independent solutions of this problem does not exceed $(p-q)^2$.

On the other hand, the functions

$$i(z\bar{z}-1)^q z^k \bar{z}^k \quad (k=0,\dots,p-q-1);$$
$$i(z\bar{z}-1)^q(z^k\bar{z}^l+\bar{z}^k z^l), \quad (z\bar{z}-1)^q(z^k\bar{z}^l-\bar{z}^k z^l),$$
$$l=0,\dots,k-1, \quad k=1,2,\dots,p-q-1,$$

are linearly independent solutions of the homogeneous problem (6.51) – (6.53) and their number is equal to $(p-q)^2$. Therefore, the number of linearly independent solutions of this problem is equal to $(p-q)^2$ and the general solution is defined by (6.54).

Theorems 6.4 and 6.5 are proved.

For unique solvability of the above considered problem let us take the supplementary conditions

$$\frac{\partial^{k+l}w(x_0,y_0)}{\partial z^k\partial\bar{z}^l}=d_{kl}, \quad l=0,\dots,k-1;\ k=1,\dots,p-q-1, \tag{6.66}$$

$$Im\frac{\partial^{2k}w(x_0,y_0)}{\partial z^k\partial\bar{z}^l}=d_k, \quad k=0,1,\dots,n-1, \tag{6.67}$$

where $w(x,y)=(z\bar{z}-1)^{-q}u(x,y)$ and d_{kl} are complex, while d_k are real constants, (x_0,y_0) is a fixed point in the domain D.

The problem (6.51) – (6.53) with the supplementary conditions (6.66), (6.67) is uniquely solvable. This statement follows directly from Theorem 6.4 and formula (6.54).

6.5 Riemann–Hilbert Problem for Second-Order Improperly Elliptic Equations in Simple Connected Domains

Let D be a simple connected bounded domain with boundary Γ. We consider the improperly elliptic equation

$$L_2u\equiv\frac{\partial^2u}{\partial y^2}+A_1\frac{\partial^2u}{\partial x\partial y}+A_2\frac{\partial^2u}{\partial x^2}=0, \quad (x,y)\in D, \tag{6.68}$$

where A_1 and A_2 are complex constants.

Boundary condition for Riemann–Hilbert problem for equation (6.68) has the following form:

$$Re u(x,y)=f_0(x,y), \quad (x,y)\in\Gamma, \tag{6.69}$$

$$Re\frac{\partial u(x,y)}{\partial N}=f_1(x,y), \quad (x,y)\in\Gamma, \tag{6.70}$$

where N is the exterior normal to the boundary Γ at the point (x, y) and $f_k(x, y)$ $(k = 0, 1)$ are given real-valued functions.

The results of the present section are used in section 6.7 for the efficient solution of the problem (6.68) – (6.70) in the circle.

Let λ_1 and λ_2 be the roots of the characteristic equation

$$\lambda^2 + A_1\lambda + A_2 = 0. \tag{6.71}$$

Since equation (6.68) is improperly elliptic, then without loss of generality we can assume that

$$Im\lambda_k > 0, \quad k = 1, 2. \tag{6.72}$$

Let, furthermore, $a_j = ReA_j$, $b_j = ImA_j$, $j = 1, 2$ and

$$\alpha = \frac{-b_2}{a_1 b_2 - a_2 b_1}, \quad \beta = \frac{b_1}{a_1 b_2 - a_2 b_1}. \tag{6.73}$$

Since λ_1 and λ_2 are the roots of equation (6.71), then

$$\begin{aligned} &A_1 = a_1 + ib_1 = -(\lambda_1 + \lambda_2), \quad A_2 = a_2 + ib_2 = \lambda_1\lambda_2, \\ &a_1 b_2 - a_2 b_1 = -|\lambda_1|^2 Im\lambda_2 - |\lambda_2|^2 Im\lambda_1. \end{aligned} \tag{6.74}$$

From (6.72) and (6.74) we obtain

$$a_1 b_2 - a_2 b_1 \neq 0.$$

The following theorem takes place:

Theorem 6.6 *The non-homogeneous problem (6.68) – (6.70) is always solvable and the general solution of the corresponding homogeneous problem is defined by*

$$u(x, y) = ic_1 + ic_2 x + ic_3 y + ic_4(y^2 + 2\alpha xy + \beta x^2), \tag{6.75}$$

where α and β are defined by (6.73) and c_j $(j = 1, \ldots, 4)$ are arbitrary real constants.

Proof. Together with the operator L_2 (cf. (6.68)), we consider the operator

$$L_2^* u \equiv \frac{\partial^2 u}{\partial y^2} + \overline{A}_1 \frac{\partial^2 u}{\partial x \partial y} + \overline{A}_2 \frac{\partial^2 u}{\partial x^2},$$

where $\overline{A}_1$ and $\overline{A}_2$ are the complex conjugate numbers to A_1 and A_2, respectively.

Let us consider the equation

$$L_2 L_2^* v = 0, \quad (x, y) \in D. \tag{6.76}$$

The characterestic equation corresponding to (6.76) is

$$(\lambda^2 + A_1\lambda + A_2)(\lambda^2 + \overline{A}_1\lambda + \overline{A}_2) = 0.$$

Therefore (6.76) is a 4-th order elliptic equation with real coefficients. Hence the Dirichlet problem for (6.76) in the domain D is uniquely solvable (see [26]). Dirichlet boundary conditions for this equation have the form

$$v(x,y) = f_0(x,y), \quad (x,y) \in \Gamma, \tag{6.77}$$

$$\frac{\partial v(x,y)}{\partial N} = f_1(x,y), \quad (x,y) \in \Gamma, \tag{6.78}$$

where $f_0(x,y)$ and $f_1(x,y)$ are the same functions as in the conditions (6.69) and (6.70).

If $u(x,y)$ is a solution of equation (6.68), then $L_2^*\overline{u(x,y)} = 0$ and the function

$$v(x,y) = \frac{1}{2}(u(x,y) + \overline{u(x,y)}) \equiv Re u(x,y) \tag{6.79}$$

is a solution of equation (6.76).

The inverse is also true: if $v(x,y)$ is a solution of equation (6.76), then it can be represented in the form (6.79), where $u(x,y)$ is a solution of equation (6.68). This follows directly from the formula of the general solution of (6.76) (see [11]).

Let $v(x,y)$ be a solution of the problem (6.76) – (6.78). Let us represent $v(x,y)$ in the form (6.79), where $u(x,y)$ is a solution of equation (6.68). Substituting $v(x,y)$ from (6.79) into (6.77) and (6.78) we get the conditions (6.69), (6.70). Thus, $u(x,y)$ is a solution of the problem (6.68) – (6.70). Therefore the non-homogeneous problem (6.68) – (6.70) is always solvable.

Let us prove that the homogeneous problem (6.68) – (6.70) (with $f_0 = f_1 \equiv 0$) has exactly four linearly independent solutions. Indeed, if $u(x,y)$ is a solution the problem (6.76) – (6.78), then $v(x,y) = Re u(x,y)$ is a solution of the homogeneous problem (6.76) – (6.78) (with $f_0 = f_1 \equiv 0$).

However, since the homogeneous problem (6.76) – (6.78) has only zero solution, then $Re u(x,y) \equiv 0$, i.e.

$$u(x,y) = iw(x,y), \tag{6.80}$$

where $w(x,y)$ is a real-valued solution of equation (6.68).

On the other hand, any function of type (6.80) satisfies the boundary conditions (6.69) and (6.70). Thus, the problem of solution of the homogeneous problem (6.68) – (6.70) is reduced to the construction of real-valued solutions of equation (6.68).

Now let us find linearly independent real solutions of equation (6.68). Let us consider two cases.

Case I. Let $\lambda_1 \neq \lambda_2$. Then the general solution $w(x,y)$ of equation (6.68) is defined by (see [16])

$$w(x,y) = \varphi_1(x+\lambda_1 y) + \varphi_2(x+\lambda_2 y), \tag{6.81}$$

where $\varphi_j(x+\lambda_j y)$ is an arbitrary analytic with respect to argument $x+\lambda_j y$ function at $(x,y) \in D$, $(j=1,2)$.

Without loss of generality, we can assume that $(0,0) \in D$. The function $\varphi_j(x+\lambda_j y)$ $(j=1,2)$ in the neighborhood of the point $(0,0)$ can be expanded into Taylor series

$$\varphi_j(x+\lambda_j y) = \sum_{k=0}^{\infty} B_{jk}(x+\lambda_j y)^k \quad for \quad |\, z \,| \leq \varepsilon, \tag{6.82}$$

where $z = x+iy$ and ε is a sufficiently small positive number.

Substituting $\varphi_j(x+\lambda_j y)$ $(j=1,2)$ from (6.82) into (6.81), we get

$$w(x,y) = \sum_{k=0}^{\infty}\sum_{j=1}^{2} B_{jk}(x+\lambda_j y)^k, \quad |\, z \,| \leq \varepsilon. \tag{6.83}$$

Let $w(x,y)$ be a real-valued solution of equation (6.68). Then

$$w(x,y) - \overline{w(x,y)} = 0. \tag{6.84}$$

Substituting $w(x,y)$ from (6.83) into (6.84), for $k=0,1,2,\ldots$ we get

$$B_{1k}(x+\lambda_1 y)^k + B_{2k}(x+\lambda_2 y)^k - \overline{B}_{1k}(x+\overline{\lambda}_1 y) - \overline{B}_{2k}(x+\overline{\lambda}_2 y)^k \equiv 0. \tag{6.85}$$

From (6.85) we obtain

$$B_{1k}\lambda_1^j + B_{2k}\lambda_2^j - \overline{B}_{1k}\overline{\lambda}_1^j - \overline{B}_{2k}\overline{\lambda}_2^j = 0, \quad j=0,\ldots,k. \tag{6.86}$$

Let Δ_4 be the Vandermond determinant, whose columns are the vectors

$$(1,\lambda_1,\lambda_1^2,\lambda_1^3),\ (1,\lambda_2,\lambda_2^2,\lambda_2^3),\ (1,\overline{\lambda}_1,\overline{\lambda}_1^2,\overline{\lambda}_1^3),\ (1,\overline{\lambda}_2,\overline{\lambda}_2^2,\overline{\lambda}_2^3).$$

Since $Im\lambda_j > 0$ $(j=1,2)$ and $\lambda_1 \neq \lambda_2$, then

$$\Delta_4 \neq 0. \tag{6.87}$$

Therefore from (6.86) we obtain

$$B_{1k} = 0, \quad B_{2k} = 0 \quad \text{for} \quad k \geq 3. \tag{6.88}$$

From this and (6.83) it follows that

$$w(x,y) = c_1 + c_2x + c_3y + c_4y^2 + 2c_5xy + c_6x^2 \quad \text{for} \quad |z| < \varepsilon, \tag{6.89}$$

where c_j ($j = 1, \ldots, 6$) are some constants. Since $w(x,y)$ is a real-valued solution of equation (6.68), the coefficients c_j ($j = 1, \ldots, 6$) of the polynomial (6.89) are real.

Since both sides of (6.89) satisfy the elliptic equation (6.68) in the domain D and are equal in the neighborhood of the point $(0,0)$, they coincide in D, i.e.

$$w(x,y) = c_1 + c_2x + c_3y + c_4y^2 + 2c_5xy + c_6x^2, \quad (x,y) \in D. \tag{6.90}$$

Substituting $u = w(x,y)$ from (6.90) into equation (6.68), we get

$$c_4 + A_1c_5 + A_2c_6 = 0.$$

Since c_4, c_5 and c_6 are real, this equation can be written in the form

$$c_4 + a_1c_5 + a_2c_6 = 0, \tag{6.91}$$

$$b_1c_5 + b_2c_6 = 0, \tag{6.92}$$

where $a_j = ReA_j$, $b_j = ImA_j$ ($j = 1; 2$).

Solving the system (6.91), (6.92) with respect to c_5 and c_6, we obtain

$$c_5 = \alpha c_4, \quad c_6 = \beta c_4, \tag{6.93}$$

where α and β are defined by (6.73).

Substituting c_5 and c_6 from (6.93) into (6.90), we get

$$w(x,y) = c_1 + c_2x + c_3y + c_4(y^2 + 2\alpha xy + \beta y^2). \tag{6.94}$$

Thus, the real-valued solutions of equation (6.68) are defined by (6.94), where c_j ($j = 1, \ldots, 4$) are arbitrary real constants. From (6.80) and (6.94) follows (6.75) for $\lambda_1 \neq \lambda_2$.

Case II. Let $\lambda_1 = \lambda_2$, $Im\lambda_1 > 0$. Then the general solution of (6.68) is defined by (see [16])

$$u(x,y) = \varphi_1(x + \lambda_1 y) + y\varphi_2(x + \lambda_2 y), \tag{6.95}$$

where $\varphi_j(x + \lambda_j y)$ is an arbitrary analytic with respect to $x + \lambda_j y$, function at $(x,y) \in D$ ($j = 1, 2$). Using (6.95) and the above arguments we can prove (6.75).

Theorem 6.6 is proved.

Let us now indicate additional conditions which provide the unique solvability of the problem (6.68) – (6.70). Based on (6.75) the additional conditions are taken in the form

$$Im u(x_0, y_0) = d_0, \quad Im \frac{\partial u(x_0, y_0)}{\partial x} = d_1,$$
$$Im \frac{\partial u(x_0, y_0)}{\partial y} = d_2, \quad Im \frac{\partial^2 u(x_0, y_0)}{\partial y^2} = d_3, \tag{6.96}$$

where d_0, d_1, d_2, d_3 are given real numbers and (x_0, y_0) is a fixed point in the closed domain $D + \Gamma$.

From Theorem 6.6 it follows that the problem (6.68) – (6.70) with additional condition (6.96) is uniquely solvable.

Now we take the boundary conditions for equation (6.68) in the form

$$Re \frac{\partial u(x, y)}{\partial x} = g_0(x, y), \quad (x, y) \in \Gamma, \tag{6.97}$$

$$Re \frac{\partial u(x, y)}{\partial y} = g_1(x, y), \quad (x, y) \in \Gamma, \tag{6.98}$$

where $g_0(x, y)$ and $g_1(x, y)$ are given functions on Γ. The following theorem takes place.

Theorem 6.7 *The non-homogeneous problem (6.68), (6.97), (6.98) has a solution if and only if the functions $g_0(x, y)$ and $g_1(x, y)$ satisfy the condition*

$$\int_\Gamma g_0(x, y)dx + g_1(x, y)dy = 0. \tag{6.99}$$

The general solution of the homogeneous problem (6.68), (6.97), (6.98) (with $f_0 = f_1 \equiv 0$) is defined by

$$u(x, y) = c_0 + ic_1 + ic_2 x + ic_3 y + ic_4(y^2 + 2\alpha xy + \beta x^2),$$

where α and β are the constants from (6.73) and c_j $(j = 0, \ldots, 4)$ are arbitrary real constants.

Proof. Let the problem (6.68), (6.97), (6.98) have a solution $u(x, y)$. Then from (6.97) and (6.98) it follows that

$$\int_\Gamma g_0(x, y)dx + g_1(x, y)dy = Re \left[\int_\Gamma \frac{\partial u}{\partial x}dx + \frac{\partial u}{\partial y}dy \right] = Re \int_\Gamma du = 0.$$

Hence, the condition (6.99) is necessary for solvability of the problem (6.68), (6.97), (6.98).

Next, assuming (6.99) we prove that the problem (6.68), (6.97), (6.98) is solvable.

The conditions (6.97) and (6.98) can be written as follows

$$Re\frac{\partial u(x,y)}{\partial s} = g_2(x,y), \quad (x,y) \in \Gamma, \tag{6.100}$$

$$Re\frac{\partial u(x,y)}{\partial N} = g_3(x,y), \quad (x,y) \in \Gamma, \tag{6.101}$$

where $\frac{\partial u}{\partial N}$ and $\frac{\partial u}{\partial s}$ are derivatives with respect to the exterior normal and arc of the curve Γ, respectively and

$$g_2(x,y) = -g_0(x,y)\cos(\widehat{N,y}) + g_1(x,y)\cos(\widehat{N,x}),$$
$$g_3(x,y) = g_0(x,y)\cos(\widehat{N,x}) + g_1(x,y)\cos(\widehat{N,y}),$$

N is the exterior normal at $(x,y) \in \Gamma$, $(\widehat{N,x})$ and $(\widehat{N,y})$ are the angles between N and positive axes X and Y, respectively.

Integrating both sides of (6.100) with respect to s, we obtain

$$Reu(x,y) = g_4(x,y) + c_0, \quad (x,y) \in \Gamma, \tag{6.102}$$

where

$$g_4(x,y) = \int_{(x_0,y_0)}^{(x,y)} g_2(x,y)ds \equiv \int_{(x_0,y_0)}^{(x,y)} g_0(x,y)dx + g_1(x,y)dy,$$

c_0 is an arbitrary real constant and (x_0,y_0) is a fixed point on Γ. The integration is realized along the arc of the contour Γ from (x_0,y_0) to (x,y). From (6.99) it follows that the function $g_4(x,y)$ is continuous on Γ.

Thus, the conditions (6.97), (6.98) are equivalent to (6.101), (6.102). Hence, according to Theorem 6.6, the problem (6.68), (6.97), (6.98) is solvable.

Let us write the general solution of the problem (6.68), (6.97), (6.98) (at $g_0 = g_1 = 0$). The conditions (6.101) and (6.102) are rewritten as

$$Re\frac{\partial u(x,y)}{\partial N} = 0, \quad (x,y) \in \Gamma, \tag{6.103}$$

$$Reu(x,y) = c_0, \quad (x,y) \in \Gamma. \tag{6.104}$$

Setting

$$w(x,y) = u(x,y) - c_0. \tag{6.105}$$

in (6.68), (6.103), (6.104), we get the homogeneous problem (6.68) – (6.70) with respect to $w(x, y)$.

According to (6.75), we have

$$w(x, y) = ic_1 + ic_2x + ic_3y + ic_4(y^2 + 2\alpha xy + \beta x^2). \tag{6.106}$$

From (6.105) and (6.106) we obtain the desired formula for the general solution of the corresponding homogeneous problem.

Theorem 6.7 is proved.

To determine a unique solution of the considered problem, we take the additional conditions

$$u(x_0, y_0) = a_0, \quad Im\frac{\partial u(x_0, y_0)}{\partial x} = a_1,$$
$$Im\frac{\partial u(x_0, y_0)}{\partial y} = a_2, \quad Im\frac{\partial^2 u(x_0, y_0)}{\partial y^2} = a_3,$$

where a_0 is a given complex, a_1, a_2 and a_3 are given real numbers and (x_0, y_0) is a fixed point in the closed domain $D \bigcup \Gamma$.

6.6 Dirichlet Type Problem for Third-Order Improperly Elliptic Equations

Let D be a simple connected bounded domain with boundary Γ. In D we consider the third-order elliptic equation:

$$\sum_{k=0}^{3} A_k \frac{\partial^3 u}{\partial x^k \partial y^{3-k}} = 0, \quad (x, y) \in D, \tag{6.107}$$

where A_k are complex constants.

Let λ_1, λ_2 and λ_3 be the roots of the characteristic equation

$$A_0\lambda^3 + A_1\lambda^2 + A_2\lambda + A_3 = 0.$$

We assume that

$$Re\lambda_1 > 0, \quad Re\lambda_2 > 0, \quad \lambda_3 = \overline{\lambda_1}.$$

The boundary condition for equation (6.107) is taken in the form:

$$u(x, y) = f_0(x, y), \quad (x, y) \in \Gamma, \tag{6.108}$$

$$Re\frac{\partial u(x, y)}{\partial N} = f_1(x, y), \quad (x, y) \in \Gamma, \tag{6.109}$$

where $f_0(x,y)$ and $f_1(x,y)$ are given functions on Γ, N is the exterior normal to the boundary Γ at the point $(x,y) \in \Gamma$ ($f_1(x,y)$ is a real-valued).

Together with the problem (6.107) – (6.109) consider the following Dirichlet problem:

$$\frac{\partial^2 w}{\partial y^2} - (\lambda_1 + \overline{\lambda}_1)\frac{\partial^2 w}{\partial x \partial y} + |\lambda_1|^2 \frac{\partial^2 w}{\partial x^2} = 1 \quad (x,y) \in D, \tag{6.110}$$

$$w(x,y) = 0, \quad (x,y) \in \Gamma. \tag{6.111}$$

Since the coefficients of the second order elliptic equation (6.110) are real, the problem (6.110), (6.111) is uniquely solvable.

The following theorem takes place.

Theorem 6.8 *The non-homogeneous problem (6.107) – (6.109) is always solvable and the general solution of the corresponding homogeneous problem is defined by*

$$u(x,y) = icw(x,y), \tag{6.112}$$

where $w(x,y)$ is a solution of the problem (6.110), (6.111), and c is an arbitrary real constant.

Proof. We consider the following cases.

Case I. Let $\lambda_1 \neq \lambda_2$. Then the general solution of equation (6.107) is defined by (see [16])

$$u(x,y) = \varphi_1(x + \lambda_1 y) + \varphi_2(x + \lambda_2 y) + \overline{\varphi_3(x + \lambda_1 y)}, \tag{6.113}$$

where $\varphi_1(x+\lambda_1 y), \varphi_2(x+\lambda_2 y)$ and $\varphi_3(x+\lambda_1 y)$ are arbitrary analytic functions on arguments $x + \lambda_1 y, x + \lambda_2 y$ and $x + \lambda_1 y$, respectively.

The boundary condition (6.108) will be written as

$$Reu(x,y) = g_0(x,y), \quad (x,y) \in \Gamma, \tag{6.114}$$

$$Imu(x,y) = g_1(x,y), \quad (x,y) \in \Gamma, \tag{6.115}$$

where $g_0(x,y) = Ref_0(x,y)$ and $g_1(x,y) = Imf_0(x,y)$.

Substituting the general solution (6.113) into the boundary conditions (6.114) and (6.109) we get

$$Rev(x,y) = g_0(x,y), \quad (x,y) \in \Gamma, \tag{6.116}$$

$$Re\frac{\partial v(x,y)}{\partial N} = f_1(x,y), \quad (x,y) \in \Gamma, \tag{6.117}$$

where

$$v(x,y) = \varphi_1(x+\lambda_1 y) + \varphi_2(x+\lambda_2 y) + \varphi_3(x+\lambda_1 y). \tag{6.118}$$

It is easy to check that $v(x,y)$ is a solution of

$$\frac{\partial^2 v}{\partial y^2} - (\lambda_1+\lambda_2)\frac{\partial^2 v}{\partial x \partial y} + \lambda_1\lambda_2 \frac{\partial^2 v}{\partial x^2} = 0. \tag{6.119}$$

Observe that

$$\lambda^2 - (\lambda_1+\lambda_2)\lambda + \lambda_1\lambda_2 = 0 \tag{6.120}$$

is the characteristic equation for (6.119).

The roots of the characteristic equation (6.120) are $\lambda = \lambda_1$ and $\lambda = \lambda_2$. Since $Im\lambda_1 > 0, Im\lambda_2 > 0$, then according to Theorem 6.6, the problem (6.116), (6.117), (6.119) is solvable. Let $v(x,y) = v_0(x,y)$ be a particular solution of this problem. Substituting $v(x,y) = v_0(x,y)$ into (6.118) we obtain

$$\varphi_1(x+\lambda_1 y) + \varphi_2(x+\lambda_2 y) = v_0(x,y) - \varphi_3(x+\lambda_1 y). \tag{6.121}$$

From (6.113) and (6.121), excluding $\varphi_1(x+\lambda_1 y) + \varphi_2(x+\lambda_2 y)$, we get

$$u(x,y) = v_0(x,y) - \varphi_3(x+\lambda_1 y) + \overline{\varphi_3(x+\lambda_1 y)}. \tag{6.122}$$

Substituting $u(x,y)$ from (6.122) into the boundary conditions (6.115) we have

$$Im\varphi_3(x+\lambda_1 y) = \frac{1}{2}(Imv_0(x,y) - g_1(x,y)), \quad (x,y) \in \Gamma. \tag{6.123}$$

It is known (see [15]), that the boundary problem (6.123) with respect to analytic function $\varphi_3(x+\lambda_1 y)$ on argument $x+\lambda_1 y$ in simple connected domains is always solvable and the solution is defined within a real additive constant.

Since λ_1 and λ_2 are the solutions of the characteristic equation (6.120), any solution of the equation (6.119) can be represented in the following form (see [16]):

$$v(x,y) = \psi_1(x+\lambda_1 y) + \psi_2(x+\lambda_2 y),$$

where $\psi_1(x+\lambda_1 y)$ and $\psi_2(x+\lambda_2 y)$ are analytic functions on arguments $x+\lambda_1 y$ and $x+\lambda_2 y$, respectively. From this it follows that any solution of (6.119) is a solution of equation (6.107) (the converse is not the case).

Hence, the function (6.122) is a solution of equation (6.107). Thus, a particular solution of the problem (6.107) – (6.109) can be defined by (6.122), where $v_0(x,y)$ is a particular solution of the problem (6.116), (6.117), (6.119) and $\psi_3(x+\lambda_1 y)$ is a particular solution of the problem (6.123). Therefore, the non-homogeneous problem (6.107) – (6.109) is always solvable.

Now consider the corresponding homogeneous problem ($f_0 = f_1 = 0$). According to Theorem 6.6, the general solution of the homogeneous problem (6.116), (6.117), (6.119) (with $g_0 = f_1 = 0$) is defined by

$$v(x,y) = ic_1 + ic_2 x + ic_3 y + ic_4(y^2 + 2\alpha xy + \beta x^2), \tag{6.124}$$

where α and β are defined by (6.73) ($A_1 = -\lambda_1 - \lambda_2,\ A_2 = \lambda_1\lambda_2$), and c_j ($j = 1, \ldots, 4$) are arbitrary real constants.

From (6.113) and (6.118) we have

$$u(x,y) = \overline{\varphi_3(x+\lambda_1 y)} - \varphi_3(x+\lambda_1 y) + v(x,y), \tag{6.125}$$

where $v(x,y)$ is defined by (6.124). It is easy to check that the function $u(x,y)$ defined by (6.125) satisfies

$$\frac{\partial^2 u}{\partial y^2} - (\lambda_1 + \overline{\lambda_1})\frac{\partial^2 u}{\partial x \partial y} + \mid \lambda_1 \mid^2 \frac{\partial^2 u}{\partial x^2} = 2ic_4(1 - \alpha(\lambda_1 + \overline{\lambda_1}) + \beta \mid \lambda_1 \mid^2). \tag{6.126}$$

Therefore, a solution $u(x,y)$ of the homogeneous problem (6.107) – (6.109) satisfies equation (6.126) and the boundary condition

$$u(x,y) = 0, \quad (x,y) \in \Gamma. \tag{6.127}$$

Since the coefficients of the elliptic equation (6.126) are real, the problem (6.126), (6.127) is uniquely solvable (see [26]). It is obvious that this solution has the form

$$u(x,y) = 2ic_4(1 - \alpha(\lambda_1 + \overline{\lambda_1}) + \beta \mid \lambda_1 \mid^2) w(x,y), \tag{6.128}$$

where $w(x,y)$ is a solution of the problem (6.110), (6.111).

Hence, a solution of the homogeneous problem (6.107) – (6.109) is represented in the form (6.112).

The converse assertion is also valid: any function of the form (6.112) is a solution of the homogeneous problem (6.107) – (6.109).

Indeed, applying the operator $\frac{\partial}{\partial y} - \lambda_2 \frac{\partial}{\partial x}$ to both sides of equation (6.110), we obtain

$$(\frac{\partial}{\partial y} - \lambda_2 \frac{\partial}{\partial x})(\frac{\partial}{\partial y} - \lambda_1 \frac{\partial}{\partial x})(\frac{\partial}{\partial y} - \overline{\lambda_1}\frac{\partial}{\partial x})u = 0. \tag{6.129}$$

Since the roots of equation $A_0\lambda^3 + A_1\lambda^2 + A_2\lambda + A_3 = 0$ are $\lambda_1,\ \lambda_2$ and $\overline{\lambda_1}$, equations (6.107) and (6.129) coincide. Therefore, the function $u = ic_4 w(x,y)$ satisfies equations (6.107).

Since $w(x,y)$ is a real-valued function and satisfies the condition (6.111), the function $u(x,y) = ic_4 w(x,y)$ satisfies (6.108) and (6.109) when $f_0 = f_1 = 0$.

Hence, the general solution of the homogeneous problem (6.107) – (6.109) is defined by (6.112).

Theorem 6.8 in the case $\lambda_1 \neq \lambda_2$ is proved.

Case II. Let $\lambda_1 = \lambda_2$. Then the general solution of (6.107) is defined by (see [16]):

$$u(x, y) = \varphi_1(x + \lambda_1 y) + y\varphi_2(x + \lambda_1 y) + \overline{\varphi_3(x + \lambda_1 y)},$$

where $\varphi_j(x + \lambda_1 y)$ $(j = 1, 2, 3)$ are arbitrary analytic on argument $x + \lambda_1 y$ functions. Using this formula, Theorem 6.8 in this case can be proved as in the previous case.

Theorem 6.8 is proved.

Now let us indicate supplementary conditions providing the unique solvability of the problem (6.107) – (6.109).

To this end, we show that at any point of the domain D a solution of the problem (6.110) – (6.111) is different from zero.

Let (x_0, y_0) be a fixed point from D. Substituting the function

$$w(x, y) = v(x, y) + \frac{1}{2(1+ \mid \lambda_1 \mid^2)}[(x - x_0)^2 + (y - y_0)^2], \tag{6.130}$$

into (6.110), (6.111), we obtain

$$\frac{\partial^2 v}{\partial y^2} - (\lambda_1 + \overline{\lambda_1})\frac{\partial^2 v}{\partial x \partial y} + |\lambda_1|^2 \frac{\partial^2 v}{\partial x^2} = 0, \quad (x, y) \in D, \tag{6.131}$$

$$v(x, y) = -\frac{1}{2(1+ \mid \lambda_1 \mid^2)}[(x - x_0)^2 + (y - y_0)^2], \quad (x, y) \in \Gamma. \tag{6.132}$$

It is known (see [25]), that a solution of equation (6.131) takes its supremum on the boundary of the domain. From the condition (6.132) it follows that the maximum of a function $v(x, y)$ on the boundary of D is negative.

We have

$$v(x, y) < 0, \quad (x, y) \in D. \tag{6.133}$$

From (6.130) and (6.133) we have

$$w(x_0, y_0) = v(x_0, y_0) < 0. \tag{6.134}$$

In view of (6.112) and (6.134) as the additional condition we take

$$Imu(x_0, y_0) = a_0, \tag{6.135}$$

where a_0 is a given real number and (x_0, y_0) is a fixed point from D.

It follows from Theorem 6.8 and (6.134), that the problem (6.107) – (6.109) with additional condition (6.135) is uniquely solvable.

6.7 Riemann–Hilbert Problem for Second-Order Improperly Elliptic Equations in the Circle

Let D be the unit circle $x^2 + y^2 < 1$ and Γ be the circumference $x^2 + y^2 = 1$. In this section we indicate efficient methods of solution of the problem (6.68) – (6.70) and (6.68), (6.97), (6.98) in the circle.

To simplify the construction we will assume that the roots λ_1 and λ_2 of the characteristic equation (6.71) satisfy the conditions

$$Im\lambda_1 > 0, \quad Im\lambda_2 > 0, \quad \lambda_1 \neq \lambda_2. \tag{6.136}$$

First, let us consider the problem (6.68), (6.97), (6.98). The general solution of equation (6.68) is defined by (see [16]):

$$u(x, y) = \varphi_1(x + \lambda_1 y) + \varphi_2(x + \lambda_2 y), \tag{6.137}$$

where $\varphi_1(x + \lambda_1 y)$ and $\varphi_2(x + \lambda_2 y)$ are analytic on arguments $x + \lambda_1 y$ and $x + \lambda_2 y$ functions, when $(x, y) \in D$.

Substituting $u(x, y)$ from (6.137) into the boundary conditions (6.97) and (6.98) we get

$$Re[\psi_1(x + \lambda_1 y) + \psi_2(x + \lambda_2 y)] = g_0(x, y), \quad (x, y) \in \Gamma, \tag{6.138}$$
$$Re[\lambda_1\psi_1(x + \lambda_1 y) + \lambda_2\psi_2(x + \lambda_2 y)] = g_1(x, y), \quad (x, y) \in \Gamma, \tag{6.139}$$

where

$$\psi_j(x + \lambda_j y) = \varphi_j'(x + \lambda_j y) \quad (j = 1, 2). \tag{6.140}$$

Let D_1 and D_2 be the images of the domain D under the mappings $\zeta = x + \lambda_1 y$ and $\zeta = x + \lambda_2 y$, respectively. Since $\varphi_1(x + \lambda_1 y)$ and $\varphi_2(x + \lambda_2 y)$ are analytic functions on arguments $x + \lambda_1 y$ and $x + \lambda_2 y$ at $(x, y) \in D$, the functions $\psi_1(\zeta)$ and $\psi_2(\zeta)$ are analytic on $\zeta = \xi + i\eta$ in the domains D_1 and D_2, respectively.

Since D is the unit circle, the boundaries D_1 and D_2 are some ellipses Γ_1 and Γ_2.

Let $z = x + iy$, $\mid z \mid= 1$. Then

$$x + \lambda_j y = \nu_j(z + \frac{\mu_j}{z}), \tag{6.141}$$

where

$$\mu_j = \frac{i - \lambda_j}{i + \lambda_j}, \quad \nu_j = \frac{\lambda_j + i}{2i}, \quad j = 1, 2. \tag{6.142}$$

Since $Im\lambda_j > 0,\ j = 1, 2$, then

$$\nu_j \neq 0, \quad | \mu_j | < 1, \quad j = 1, 2. \tag{6.143}$$

Taking into account (6.141) we represent the boundary conditions (6.138) and (6.139) in the following form:

$$Re[\omega_1(z) + \omega_2(z)] = g_0(z), \quad | z | = 1, \tag{6.144}$$

$$Re[\lambda_1\omega_1(z) + \lambda_2\omega_2(z)] = g_1(z), \quad | z | = 1, \tag{6.145}$$

where

$$\omega_j(z) = \psi_j(\nu_j(z + \frac{\mu_j}{z})), \quad j = 1, 2, \tag{6.146}$$

$$g_j(z) = g_j(x, y), \quad j = 0, 1.$$

Let us consider the mapping

$$\zeta = \nu_j(z + \frac{\mu_j}{z}) \quad (j = 1, 2). \tag{6.147}$$

The function (6.147) maps conformally the ring

$$\sqrt{| \mu_j |} < | z | < 1 \tag{6.148}$$

onto the domain D_j without the segment with endpoints $-\sqrt{\mu_j}$ and $\sqrt{\mu_j}$, where $-\sqrt{\mu_j}$ and $\sqrt{\mu_j}$ are the foci of the ellipse Γ_j $(j = 1, 2)$. Hence, $\omega_j(z)$ is an analytic function in the ring (6.148). Let

$$\omega_j(z) = \sum_{k=-\infty}^{+\infty} c_{jk} z^k, \quad (j = 1, 2) \tag{6.149}$$

be Laurent series expansion of the function $\omega_j(z)$ in this ring.

From (6.146) we have

$$\omega_j(\sqrt{\mu_j} z) = \omega_j(\frac{\sqrt{\mu_j}}{z}) \quad \text{for} \quad | z | = 1, \quad j = 1, 2. \tag{6.150}$$

Substituting $\omega_j(z)$ from (6.149) into (6.150) we obtain

$$\sum_{k=-\infty}^{+\infty} c_{jk} (\sqrt{\mu_j})^k z^k = \sum_{k=-\infty}^{+\infty} c_{jk} \frac{(\sqrt{\mu_j})^k}{z^k}, \quad | z | = 1, \quad j = 1, 2. \tag{6.151}$$

From this it follows that

$$c_{jl} = \mu_j^k c_{jk} \quad \text{at} \quad l = -k, \quad k = 1, 2, \ldots, \quad j = 1, 2.$$

Hence, $\omega_j(z)$ can be rewritten as

$$\omega_j(z) = \phi_j(z) + \phi_j(\frac{\mu_j}{z}), \quad | \sqrt{\mu_j} |<| z |< 1, \quad j = 1,2, \tag{6.152}$$

where

$$\phi_j(z) = \frac{c_{j0}}{2} + \sum_{k=1}^{\infty} c_{jk} z^k \quad (j = 1,2). \tag{6.153}$$

It follows from (6.153) that the functions $\phi_1(z)$ and $\phi_2(z)$ are analytic in the circle $| z |< 1$.

Substituting $\omega_j(z)$ from (6.152) into the boundary conditions (6.144) and (6.145) to define the analytic functions $\phi_1(z)$ and $\phi_2(z)$ for $| z |= 1$ we obtain the boundary problem

$$Re\left[\phi_1(z) + \overline{\phi_1(\mu_1\overline{z})} + \phi_2(z) + \overline{\phi_2(\mu_2\overline{z})}\right] = g_0(z), \tag{6.154}$$

$$Re\left[\lambda_1\phi_1(z) + \overline{\lambda_1\phi_1(\mu_1\overline{z})} + \lambda_2\phi_2(z) + \overline{\lambda_2\phi_2(\mu_2\overline{z})}\right] = g_1(z). \tag{6.155}$$

To obtain (6.154) and (6.155) we use the following obvious relations

$$Re\phi_j(\frac{\mu_j}{z}) = Re\phi_j(\mu_j\overline{z}) \quad \text{for} \quad | z |= 1,$$
$$Re\phi_j(\mu_j\overline{z}) = Re\overline{\phi_j(\mu_j\overline{z})}.$$

Let us consider the following cases

Case I. Let

$$g_0(z) = Re[a_m z^m] \quad and \quad g_1(z) = Re[b_m z^m], \tag{6.156}$$

where $m \geq 2$ is an integer, a_m and b_m are complex constants. It is clear that the functions (6.156) satisfy the condition (6.99).

Substituting $g_0(x)$ and $g_1(z)$ from (6.156) into (6.154) and (6.155) for $| z |=$ 1, we get

$$Re[\phi_1(z) + \overline{\phi_1(\mu_1\overline{z})} + \phi_2(z) + \overline{\phi_2(\mu_2\overline{z})} - a_m z^m] = 0, \tag{6.157}$$

$$Re[\lambda_1\phi_1(z) + \overline{\lambda}_1\overline{\phi_1(\mu_1\overline{z})} + \lambda_2\phi_2(z) + \overline{\lambda}_2\overline{\phi_2(\mu_2\overline{z})} - b_m z^m] = 0. \tag{6.158}$$

Since the functions in the square brackets of (6.157) and (6.158) are analytic in the circle $| z |< 1$, from (6.157) and (6.158) we obtain

$$\phi_1(z) + \overline{\phi_1(\mu_1\overline{z})} + \phi_2(z) + \overline{\phi_2(\mu_2\overline{z})} - a_m z^m = ic, \tag{6.159}$$

$$\lambda_1\phi_1(z) + \overline{\lambda}_1\overline{\phi_1(\mu_1\overline{z})} + \lambda_2\phi_2(z) + \overline{\lambda}_2\overline{\phi_2(\mu_2\overline{z})} - b_m z^m = id, \tag{6.160}$$

where c and d are arbitrary real constants. Substituting $\phi_j(z)$ from (6.153) into (6.159) and (6.160) to define the coefficients c_{jk} we obtain the following systems of equations

$$\begin{cases} c_{10} + \bar{c}_{10} + c_{20} + \bar{c}_{20} = 2ic, \\ \lambda_1 c_{10} + \overline{\lambda}_1 \bar{c}_{10} + \lambda_2 c_{20} + \overline{\lambda}_2 \bar{c}_{20} = 2id. \end{cases} \tag{6.161}$$

$$\begin{cases} c_{1k} + \bar{c}_{1k}\overline{\mu}_1^k + c_{2k} + \bar{c}_{2k}\overline{\mu}_2^k = 0, \ k \neq m, \ k = 1, 2, \ldots \\ \lambda_1 c_{1k} + \overline{\lambda}_1 \bar{c}_{1k}\overline{\mu}_1^k + \lambda_2 c_{2k} + \overline{\lambda_2 c_{2k}\overline{\mu}_2^k} = 0. \end{cases} \tag{6.162}$$

$$\begin{cases} c_{1m} + \bar{c}_{1m}\overline{\mu}_1^m + c_{2m} + \bar{c}_{2m}\overline{\mu}_2^m = a_m, \\ \lambda_1 c_{1m} + \overline{\lambda}_1 \bar{c}_{1m}\overline{\mu}_1^m + \lambda_2 c_{2m} + \overline{\lambda}_2 \bar{c}_{2m}\overline{\mu}_2^m = b_m. \end{cases} \tag{6.163}$$

According to Theorem 6.7, the problem (6.68), (6.97), (6.98) with the boundary conditions (6.156) is solvable. Hence, for any complex values of a_m and b_m the system (6.163) is solvable with respect to c_{1m} and c_{2m}. From Lemma 6.4 it follows that the system (6.163) is uniquely solvable. Setting $c = d = 0$ in (6.161), as a solution of the system (6.161) and (6.162) we can take

$$c_{1k} = 0, \quad c_{2k} = 0, \quad k \neq m, \quad k = 0, 1, 2, \ldots. \tag{6.164}$$

Therefore

$$\phi_1(z) = c_{1m} z^m, \quad \phi_2(z) = c_{2m} z^m, \tag{6.165}$$

where c_{1m} and c_{2m} satisfy the system (6.163).

Case II. Let the functions $g_0(z)$ and $g_1(z)$ be represented in the form

$$g_0(z) = Re \sum_{k=2}^{\infty} a_m z^m, \quad g_1(z) = Re \sum_{k=0}^{\infty} b_m z^m. \tag{6.166}$$

Then it is obvious that a particular solution of the boundary value problem (6.154), (6.155) is defined by

$$\phi_1(z) = \sum_{k=2}^{\infty} c_{1m} z^m \quad \text{and} \quad \phi_2(z) = \sum_{k=2}^{\infty} c_{2m} z^m, \tag{6.167}$$

where c_{1m} and c_{2m} $(m = 2, 3, \ldots)$ satisfy the system (6.163). We have shown above that for $m \geq 2$ the system (6.163) is uniquely solvable.

From (6.140) and (6.146) we have

$$\varphi_j(x + \lambda_j y) = \int_0^{x+\lambda_j y} \psi_j(\zeta) d\zeta + a_j \quad (j = 1, 2), \tag{6.168}$$

$$\psi_j(\zeta) = \omega_j(\delta_j(\zeta)), \quad (j = 1, 2), \tag{6.169}$$

where a_j is an arbitrary constant and

$$\delta_j(\zeta) = \frac{\zeta + \sqrt{\zeta^2 - 4\nu_j^2\mu_j}}{2\nu_j}.$$

By $\sqrt{\zeta^2 - 4\nu_j^2\mu_j}$ we mean the branch of root which is continuous on the whole complex plane without the segment with endpoints $-2\nu_j\sqrt{\mu_j}$ and $2\nu_j\sqrt{\mu_j}$ and satisfies the condition

$$\lim_{|\zeta|\to+\infty} \frac{\sqrt{\zeta^2 - 4\nu_j^2\mu_j}}{\zeta} = 1.$$

Thus, if $g_0(z)$ and $g_1(z)$ have the form (6.166), a particular solution of the problem (6.68), (6.97), (6.98) is defined by (6.137), (6.168), (6.152), (6.167).

Case III. Let

$$g_0(z) = a_0, \quad g_1(z) = b_0, \tag{6.170}$$

where a_0 and b_0 are given real constants. It is easy to see that a particular solution of the problem (6.68), (6.97), (6.98) is defined by

$$u(x,y) = a_0 x + b_0 y. \tag{6.171}$$

Case IV. Let

$$g_0(z) = a_{11}x + a_{12}y, \quad \textit{and} \quad g_1(z) = a_{12}x + a_{22}y, \tag{6.172}$$

where a_{11}, a_{12} and a_{22} are given real constants.

A solution of the problem (6.68), (6.97), (6.98) in this case is represented in the form

$$u(x,y) = (\alpha_1 + i\beta_1)x^2 + 2(\alpha_2 + i\beta_2)xy + \alpha_3 y^2, \tag{6.173}$$

where α_1, α_2, α_3, β_1 and β_2 are real constants to be found.

Substituting $u(x,y)$ from (6.173) into (6.97) and (6.98) and using (6.172), we have

$$\alpha_1 = \frac{a_{11}}{2}, \quad \alpha_2 = \frac{a_{12}}{2}, \quad \alpha_3 = \frac{a_{22}}{2}. \tag{6.174}$$

Substituting $u(x,y)$ from (6.173) into (6.68), we get

$$\alpha_3 + A_1(\alpha_2 + i\beta_2) + A_2(\alpha_1 + i\beta_1) = 0. \tag{6.175}$$

Equation (6.175) can be written in the form

$$A_1\beta_2 + A_2\beta_1 = \gamma, \tag{6.176}$$

where

$$\gamma = i(\alpha_3 + A_1\alpha_2 + A_2\alpha_1) = \frac{i}{2}(a_{22} + A_1a_{12} + A_2a_{11}).$$

In section 6.5 (see (6.91)) we have shown that the equation (6.176) is uniquely solvable with respect to real constants β_1 and β_2.

Thus, in this case a particular solution of the problem (6.68), (6.97), (6.98) is defined by (6.173), where the real constants α_1, α_2, α_3, β_1 and β_2 are defined by (6.174) and (6.176).

Let us now consider the general case. Let $g_0(z)$ and $g_1(z)$ be functions defined on Γ and satisfying the condition (6.99), then these functions can be represented as a sum of functions of the form (6.166), (6.170) and (6.172).

Therefore, the general case can be reduced to the above considered cases.

Thus, we have indicated a method of construction of a particular solution of the problem (6.68), (6.97), (6.98) in the circle. The general solution of the corresponding homogeneous problem is given in section 6.5 (see Theorem 6.7).

Now we consider the problem (6.68) – (6.70) in the circle. Differentiating both sides of (6.69) along the arc s of the contour Γ, we get

$$Re(-y\frac{\partial u}{\partial x} + x\frac{\partial u}{\partial y}) = \frac{df_0(x,y)}{ds}, \quad (x,y) \in \Gamma. \tag{6.177}$$

The condition (6.70) can be written in the form

$$Re(x\frac{\partial u}{\partial x} + y\frac{\partial u}{\partial y}) = f_1(x,y), \quad (x,y) \in \Gamma. \tag{6.178}$$

Substituting $x = 1,\ y = 0$ on both sides of (6.69), we get

$$Reu(1,0) = f_0(1,0). \tag{6.179}$$

Solving (6.177) and (6.178) with respect to $Re\frac{\partial u}{\partial x}$ and $Re\frac{\partial u}{\partial y}$ we obtain

$$Re\frac{\partial u}{\partial x} = g_0(z), \quad |\, z \,|= 1, \tag{6.180}$$

$$Re\frac{\partial u}{\partial y} = g_1(z), \quad |\, z \,|= 1, \tag{6.181}$$

where

$$g_0(z) = -y\frac{df_0(x,y)}{ds} + xf_1(x,y), \quad g_1(z) = x\frac{df_0(x,y)}{ds} + yf_1(x,y). \tag{6.182}$$

From (6.182) it follows that

$$\int\limits_{\Gamma} g_0dx + g_1dy = \int\limits_{\Gamma} \frac{df_0(x,y)}{ds}ds = 0.$$

It is clear that the problem (6.68) – (6.70) is equivalent to the problem (6.68), (6.179) – (6.181).

Let $u(x, y)$ be a solution of the problem (6.68), (6.180), (6.181). Then the function $u(x, y) - u(1, 0) + f_0(1, 0)$ is a solution of the problem (6.68), (6.179) – (6.181). Hence, this function is a particular solution of the problem (6.68) – (6.70).

Thus, the construction of a particular solution of this problem is reduced to the construction of a particular solution of the previous problem. The general solution of the homogeneous problem (6.68) – (6.70) is given in section 6.5 (cf. (6.75)).

6.8 Riemann–Hilbert Problem For High Order Improperly Elliptic Equations

Let D be a simple connected bounded domain in R^2 with boundary Γ. Assume that the roots $\lambda_1, \dots, \lambda_n$ of the characteristic equation (6.2) satisfy the condition

$$Re\lambda_j > 0, \quad j = 1, \dots, n. \tag{6.183}$$

The boundary conditions for (6.1) are taken in the form

$$Re \frac{\partial^k u(x, y)}{\partial N^k} = f_k(x, y), \quad (x, y) \in \Gamma, \quad k = 0, \dots, n - 1, \tag{6.184}$$

where N is the exterior normal to the boundary Γ at the point (x, y) and $f_0(x, y), \dots, f_{n-1}(x, y)$ are given real-valued functions.

The following theorem takes place

Theorem 6.9 *If the condition (6.183) is satisfied, then the non-homogeneous problem (6.1), (6.184) is always solvable, and the corresponding homogeneous problem has exactly n^2 linearly independent solutions. The general solutions of the problem (6.1), (6.184) is defined by $u(x, y) = iw(x, y)$, where $w(x, y)$ is the general solution of (6.1) in the class of real-valued functions.*

Proof. The existence of a solution of the problem (6.1), (6.184) for $n = 2$ has been proved in section 6.5 (see Theorem 6.6). For $n > 2$ this assertion can be proved in a similar way.

In section 6.5 for $n = 2$ we have proved that the general solution of the homogeneous problem (6.1), (6.184) has the form

$$u(x, y) = iv(x, y), \tag{6.185}$$

where $v(x, y)$ is the general solution of (6.1) in the class of real-valued functions. This assertion is also true for $n > 2$ and can be proved in a similar way.

From (6.185) and Theorem 7.1 (see section 7.1) follows the second part of Theorem 6.9.

Theorem 6.9 is proved.

6.9 Dirichlet Type Problem for High Order Improperly Elliptic Equations

Let D be a simple connected domain with boundary Γ and let $\lambda_1, \ldots, \lambda_n$ be the roots of the characteristic equation (6.2). We assume that

$$Im\lambda_j > 0, \quad \text{for} \quad j = 1, \ldots, p, \tag{6.186}$$

$$Im\lambda_k < 0, \quad \text{for} \quad k = p+1, \ldots, n, \tag{6.187}$$

$$\lambda_{p+j} = \overline{\lambda_j}, \quad \text{for} \quad j = 1, \ldots, n-p, \tag{6.188}$$

where p is a positive integer such that $1 \leq p < n$. Without loss of generality we assume that $p \geq n/2$.

The boundary conditions for equation (6.1) in this case is taken in the form

$$\frac{\partial^k u(x,y)}{\partial N^k} = f_k(x,y), \quad (x,y) \in \Gamma, \quad k = 0, \ldots, q-1, \tag{6.189}$$

$$Re\frac{\partial^k u(x,y)}{\partial N^k} = f_k(x,y), \quad (x,y) \in \Gamma, \ k = q, \ldots, p-1, \tag{6.190}$$

where $q = n - p$, N is the exterior normal to the boundary Γ at the point $(x,y) \in \Gamma$, $f_0(x,y), \ldots, f_{p-1}(x,y)$ are given on Γ functions and $f_q(x,y), \ldots, f_{p-1}(x,y)$ are real-valued.

To investigate the problem (6.1), (6.189), (6.190) let us consider the following equations

$$L_p u(x,y) = 0, \quad (x,y) \in D \tag{6.191}$$

$$L_q L_q^* u(x,y) = f(x,y), \quad (x,y) \in D, \tag{6.192}$$

where

$$L_p u \equiv (\frac{\partial}{\partial x} - \lambda_1 \frac{\partial}{\partial y}) \ldots (\frac{\partial}{\partial x} - \lambda_p \frac{\partial}{\partial y}) u, \tag{6.193}$$

$$L_q u \equiv (\frac{\partial}{\partial x} - \lambda_1 \frac{\partial}{\partial y}) \ldots (\frac{\partial}{\partial x} - \lambda_q \frac{\partial}{\partial y}) u, \tag{6.194}$$

$$L_q^* u \equiv (\frac{\partial}{\partial x} - \overline{\lambda_1} \frac{\partial}{\partial y}) \ldots (\frac{\partial}{\partial x} - \overline{\lambda_q} \frac{\partial}{\partial y}) u. \tag{6.195}$$

For equation (6.192) the boundary condition is taken in the form

$$\frac{\partial^k u(x,y)}{\partial N^k} = 0, \quad (x,y) \in \Gamma, \quad k = 0, \ldots, q-1. \tag{6.196}$$

Since the coefficients of the elliptic equation (6.192) are real, for any smooth function $f(x,y)$ the problem (6.192), (6.196) is uniquely solvable.

Let Q_{pq} denote the set of functions that admit the following representation

$$v(x,y) = L_q L_q^* u, \tag{6.197}$$

where $u(x,y)$ is an arbitrary real-valued solution of equation (6.191). The set Q_{pq} is a linear space over the field of real numbers.

In section 7.1 (see Theorem 7.1) we will indicate a method of construction of real-valued solutions for equation (6.191). Substituting these solutions into (6.197) we obtain the set Q_{pq}.

In section 7.2 (see Theorem 7.4) we will also prove that the dimension of Q_{pq} is equal to $(p-q)^2$. Let $v_1(x,y), \ldots, v_m(x,y)$ $(m = (p-q)^2)$ be linearly independent functions from Q_{pq}. Denote by $u_1(x,y), \ldots, u_m(x,y)$ the solutions of the problem (6.192), (6.196), when $f(x,y)$ coincides with $v_1(x,y), \ldots, v_m(x,y)$, respectively.

The following theorem takes place

Theorem 6.10 *If the conditions (6.186) - (6.188) are satisfied, then the non-homogeneous problem (6.1), (6.189), (6.190) is always solvable, and the general solution of the corresponding homogeneous problem is defined by*

$$u(x,y) = i(c_1 u_1(x,y) + \cdots + c_m u_m(x,y)), \tag{6.198}$$

where $m = (p-q)^2$ and $c_1, \ldots, c_m$ are arbitrary real constants.

Proof. First let us construct a particular solution of the problem (6.1), (6.189), (6.190). In the paper [11] we have obtained a formula for the general solution of equation (6.1) (see also section 7.1). From this formula it follows that any solution of equation (6.1) is represented in the form

$$u(x,y) = \omega_1(x,y) + \overline{\omega_2(x,y)}, \tag{6.199}$$

where $\omega_1(x,y)$ and $\omega_2(x,y)$ are the solutions of equations

$$L_p \omega_1 = 0, \quad (x,y) \in D, \tag{6.200}$$

and

$$L_q \omega_2 = 0, \quad (x,y) \in D \tag{6.201}$$

respectively, where L_p and L_q are the operators defined by (6.193) and (6.194).

The converse assertion is also true: any function of the form (6.199) is a solution of equation (6.1).

The boundary condition (6.189) takes the form

$$Re\frac{\partial^k u(x,y)}{\partial N^k} = Ref_k(x,y), \quad (x,y) \in \Gamma, \quad k = 0,\dots,q-1, \tag{6.202}$$

$$Im\frac{\partial^k u(x,y)}{\partial N^k} = Imf_k(x,y), \quad (x,y) \in \Gamma, \quad k = 0,\dots,q-1. \tag{6.203}$$

Substituting $u(x,y)$ from (6.199) into the boundary conditions (6.190) and (6.202), we get

$$Re\frac{\partial^k \omega(x,y)}{\partial N^k} = Ref_k(x,y), \quad (x,y) \in \Gamma, \quad k = 0,\dots,p-1, \tag{6.204}$$

where

$$\omega(x,y) = \omega_1(x,y) + \omega_2(x,y). \tag{6.205}$$

It is obvious that the function $\omega(x,y)$ is a solution of the equation

$$L_p\omega = 0, \quad (x,y) \in D. \tag{6.206}$$

Thus, to determine the function $\omega(x,y)$ we need to solve the problem (6.204), (6.206). According to Theorem 6.9 (see section 6.8), this problem has a solution and its general solution is defined by

$$\omega(x,y) = \omega_0(x,y) + iw(x,y), \tag{6.207}$$

where $\omega_0(x,y)$ is a particular solution of the non-homogeneous problem (6.204), (6.206), while $w(x,y)$ is the general solution of equation (6.206) in the class of real-valued functions.

From (6.199), (6.205) and (6.207) we have

$$u(x,y) = \omega_0(x,y) - \omega_2(x,y) + \overline{\omega_2(x,y)} + iw(x,y). \tag{6.208}$$

Observe that (6.208) can be written in the form

$$u(x,y) = i\psi(x,y) + \omega_0(x,y) + iw(x,y), \tag{6.209}$$

where

$$\psi(x,y) = i(\omega_2(x,y) - \overline{\omega_2(x,y)}). \tag{6.210}$$

It is obvious that $\psi(x,y)$ is a real-valued function.

Since $L_q\omega_2(x,y) = 0$, we have $L_q^*\overline{\omega_2(x,y)} = 0$ and

$$L_qL_q^*\psi(x,y) = 0. \tag{6.211}$$

Substituting $u(x,y)$ from (6.209) into the boundary conditions (6.203), for $k = 0,\ldots,q-1$ we obtain

$$\frac{\partial^k \psi(x,y)}{\partial N^k} = Imf_k(x,y) - Im\frac{\partial^k \omega_0(x,y)}{\partial N^k} - \frac{\partial^k w(x,y)}{\partial N^k}, \quad (x,y) \in \Gamma. \quad (6.212)$$

As it has been indicated above, the problem (6.211), (6.212) is uniquely solvable with respect to the function $\psi(x,y)$. Solving this problem with $w(x,y) \equiv 0$ and substituting the obtained solution into (6.209) we get a particular solution of the problem (6.1), (6.189), (6.190).

Now let $f_k(x,y) \equiv 0$, $k = 0,\ldots,p-1$ and $\omega_0(x,y) \equiv 0$. In this case the conditions (6.212) take the following form

$$\frac{\partial^k \psi(x,y)}{\partial N^k} = -\frac{\partial^k w(x,y)}{\partial N^k}, \quad (x,y) \in \Gamma. \quad (6.213)$$

Substituting into (6.211) and (6.213) the function

$$\psi(x,y) = \varphi(x,y) - w(x,y), \quad (6.214)$$

we have

$$L_q L_q^* \varphi(x,y) = L_q L_q^* w(x,y), \quad (x,y) \in D, \quad (6.215)$$

$$\frac{\partial^k \varphi(x,y)}{\partial N^k} = 0, \quad k = 0,\ldots,q-1, \quad (x,y) \in \Gamma. \quad (6.216)$$

Substituting $\psi(x,y)$ from (6.214) into (6.209) with $\omega_0(x,y) \equiv 0$, we obtain

$$u(x,y) = i\varphi(x,y). \quad (6.217)$$

Thus, the general solution of the homogeneous problem (6.1), (6.189), (6.190) (when $f_k = 0$, $k = 1,\ldots,p-1$) is defined by (6.217), where $w(x,y)$ is a solution of equation (6.200) in the class of real-valued functions and $\varphi(x,y)$ is a solution of Dirichlet problem (6.215), (6.216).

According to the definition of functions $v_1(x,y),\ldots,v_m(x,y)$, we have

$$L_q L_q^* w(x,y) = c_1 v_1(x,y) + \ldots + c_m v_m(x,y), \quad (6.218)$$

where $c_1,\ldots,c_m$ are arbitrary real constants and $m = (p-q)^2$.

From (6.215) – (6.218) we obtain (6.198).

Theorem 6.10 is proved.

6.10 Neumann Type Problem for Improperly Elliptic Equations

Let D be the circle $x^2 + y^2 < 1$ and Γ be the circumference $x^2 + y^2 = 1$.

1. We consider equation (6.1) in this circle. Let the roots $\lambda_1, \ldots, \lambda_n$ of the characteristic equation (6.2) satisfy the condition

$$Im\lambda_j > 0, \quad j = 1, \ldots, n. \tag{6.219}$$

Boundary conditions of Neumann problem for equation (6.1) are taken in the form

$$Re\frac{\partial^k u(x,y)}{\partial N^k} = f_k(x,y), \quad k = 1, \ldots, n, \quad (x,y) \in \Gamma, \tag{6.220}$$

where N is the exterior normal to the boundary Γ at the point $(x,y) \in \Gamma$ and $f_k(x,y)$ $(k = 1, \ldots, n)$ are given on Γ real-valued functions.

Let L and L^* be the operators defined by

$$Lu \equiv \sum_{k=0}^{n} A_k \frac{\partial^n u}{\partial x^k \partial y^{n-k}} \quad \text{and} \quad L^* u \equiv \sum_{k=0}^{n} \overline{A_k} \frac{\partial^n u}{\partial x^k \partial y^{n-k}}.$$

Consider the following problem:

$$LL^*\varphi(x,y) = 0, \quad (x,y) \in D, \tag{6.221}$$

$$\frac{\partial^k \varphi(x,y)}{\partial N^k} = g_k(x,y), \quad k = 0, \ldots, n-1, \quad (x,y) \in \Gamma, \tag{6.222}$$

where

$$g_0(x,y) = f_1(x,y),$$
$$g_k(x,y) = kf_k(x,y) + f_{k+1}(x,y), \quad k = 1, \ldots, n-1,$$

and $f_1(x,y), \ldots, f_n(x,y)$ are functions from the boundary conditions (6.220). The problem (6.221), (6.222) is uniquely solvable (see [26]).

Theorem 6.11 *The non-homogeneous problem (6.1), (6.220) is solvable if and only if*

$$\varphi(0,0) = 0 \tag{6.223}$$

where $\varphi(x,y)$ is a solution of the problem (6.221), (6.222).

Theorem 6.12 *The general solution $u(x,y)$ of the homogeneous problem (6.1), (6.220) is defined by*

$$u(x,y) = c + iw(x,y) \tag{6.224}$$

where $w(x,y)$ is an arbitrary real-valued solution of equation (6.1) and c is an arbitrary real constant.

Proof of Theorems 6.11 and 6.12. Let

$$v(x,y) = x\frac{\partial u(x,y)}{\partial x} + y\frac{\partial u(x,y)}{\partial y}, \tag{6.225}$$

where $u(x,y)$ is a solution of the problem (6.1), (6.220).

Applying the operator L (see (6.1)) to the function $v(x,y)$ we obtain

$$Lv(x,y) = x\frac{\partial Lu}{\partial x} + y\frac{\partial Lu}{\partial y} + nLu, \quad (x,y) \in D. \tag{6.226}$$

Therefore, if $u(x,y)$ is a solution of equation (6.1), then the function $v(x,y)$ is a solution of this equation, too.

Let (r,θ) be the polar coordinates of a point P with Cartesian coordinates (x,y):

$$x = r\cos\theta, \quad y = r\sin\theta \tag{6.227}$$

and let $u(r,\theta)$ be the value of $u(x,y)$ at the point with polar coordinates (r,θ). The conditions (6.220) and the function $v(x,y)$ in polar coordinates can be written as follows

$$Re\frac{\partial^k u(r,\theta)}{\partial r^k} = f_k(\cos\theta, \sin\theta) \quad for \quad r = 1, \tag{6.228}$$

$$0 \le \theta \le 2\pi, \quad (k = 1, \ldots, n),$$

$$v(r,\theta) = r\frac{\partial u(r,\theta)}{\partial r}. \tag{6.229}$$

From (6.226) – (6.229) we obtain

$$Lv(x,y) = 0, \quad (x,y) \in D, \tag{6.230}$$

$$Re\frac{\partial^k v(x,y)}{\partial N^k} = g_k(x,y), \quad (x,y) \in \Gamma, \quad k = 0, \ldots, n-1, \tag{6.231}$$

where $g_k(x,y)$ $(k = 0, \ldots, n-1)$ are the same functions as in (6.222).

Substituting $x = y = 0$ into (6.225) we obtain

$$v(0,0) = 0. \tag{6.232}$$

From (6.229) we get

$$u(r,\theta) = \int_0^r \frac{v(t,\theta)}{t} dt + C(\theta), \tag{6.233}$$

where $C(\theta)$ is an arbitrary function. Changing the variable $t = r\tau$ in (6.233), we obtain

$$v(r,\theta) - \int_0^1 \frac{v(r\tau,\theta)}{\tau} d\tau = C(\theta). \tag{6.234}$$

Let us represent the integral on the left-hand side of (6.234) in Cartesian coordinates

$$\int_0^1 \frac{v(rt,\theta)}{t}dt = \int_0^1 \frac{v(xt,yt)}{t}dt. \tag{6.235}$$

Since $u(r,\theta)$ and $v(r,\theta)$ are solutions of equation (6.1), from (6.234) and (6.235) it follows that $C(\theta)$ is a solution of this equation, too. From the formula of the general solution of (6.1) it follows that any solution $w(x,y)$ of this equation in the neighborhood of the point $(0,0)$ can be expanded into the Taylor series (see [11])

$$w(x,y) = \sum_{k=0}^{\infty}\sum_{j=0}^{\infty} a_{kj}x^k y^j.$$

In polar coordinates this series has the form

$$w(r,\theta) = a_{00} + \sum_{k=1}^{\infty} C_k(\theta) r^k \quad \text{for} \quad 0 \le r \le \varepsilon, \tag{6.236}$$

where $C_k(\theta)$ are some functions depending only on θ and ε is a sufficiently small positive number.

Comparing the function $C(\theta)$ with (6.236) we obtain $C(\theta) = a_{00}$. Therefore, the equality (6.234) in Cartesian coordinates takes the form

$$u(x,y) = \int_0^1 \frac{v(xt,yt)}{t}dt + a_{00}. \tag{6.237}$$

Thus, a solution of the problem (6.1), (6.220) is defined by (6.237), where $v(x,y)$ is a solution of the problem (6.230) – (6.232).

Let

$$v(x,y) = \varphi(x,y) + i\psi(x,y), \tag{6.238}$$

where $\varphi(x,y)$ and $\psi(x,y)$ are the real and imaginary parts of a function $v(x,y)$. The conditions (6.231) and (6.232) take the form

$$\frac{\partial^k \varphi(x,y)}{\partial N^k} = g_k(x,y), \quad (x,y) \in \Gamma, \quad k = 0,\dots,n-1, \tag{6.239}$$

$$\varphi(0,0) = 0, \tag{6.240}$$

$$\psi(0,0) = 0. \tag{6.241}$$

From (6.238) we get

$$\varphi(x,y) = \frac{1}{2}(v(x,y) + \overline{v(x,y)}). \tag{6.242}$$

Since $Lv = 0$, and $L^*\overline{v} = 0$, the function $\varphi(x,y)$ is a solution of equation (6.221). From this and (6.240) it follows that (6.223) is necessary for solvability of the problem (6.1), (6.220). Let us prove its sufficiency. Let (6.223) be fulfilled. The condition (6.241) can be written in the form

$$Imv(0,0) = 0. \tag{6.243}$$

In section 6.8 it has been shown that the non-homogeneous problem (6.230), (6.231) is always solvable. Let $v_0(x,y)$ be a particular solution of this problem. Then $v(x,y) = v_0(x,y) - iImv_0(0,0)$ is also a particular solution of this problem and satisfies the condition (6.243). Substituting this solution into (6.237) with $a_{00} = 0$ we obtain a particular solution of the problem (6.1), (6.220). Therefore, the condition (6.223) is necessary and sufficient for solvability of the problem (6.1), (6.220).

Now let us construct the general solution of the homogeneous problem (6.1), (6.220). As we have shown in section 6.8, a solution of the homogeneous problem (6.230), (6.231) is defined by

$$v(x,y) = iw(x,y), \tag{6.244}$$

where $w(x,y)$ is an arbitrary real-valued solution of equation (6.230).

In section 7.1 we will prove that the number of linearly independent real-valued solutions of equation (6.230) (over the field of real numbers) is equal to n^2 and the general solution of this equation in the class of real-valued functions is defined by

$$w(x,y) = c_1 + \sum_{k=2}^{n^2} c_k w_k(x,y), \tag{6.245}$$

where $w_k(x,y)$ $(k = 2,\ldots,n^2)$ are some linearly independent homogeneous polynomials on x and y with real coefficients,

$$w_k(0,0) = 0, \quad k = 2,\ldots,n^2, \tag{6.246}$$

and c_k $(k = 1,\ldots,n^2)$ are arbitrary real constants.

From (6.244), (6.245) and (6.243) we have

$$v(x,y) = i\sum_{k=2}^{n^2} c_k w_k(x,y). \tag{6.247}$$

Substituting $v(x,y)$ from (6.247) into (6.237) we obtain

$$u(x,y) = d_0 + id_1 + \sum_{k=2}^{n^2} d_k i w_k(x,y), \tag{6.248}$$

where d_j $(j = 0, \ldots, n^2)$ are arbitrary real constants.

From (6.245) and (6.248) follows the assertion of Theorem 6.12.

In particular, from (6.248), we have

Corollary 6.1 *The homogeneous problem (6.1), (6.220) has exactly* $n^2 + 1$ *linearly independent solutions.*

2. Now let the coefficients of the elliptic equation (6.1) be real and let the order of the equation be even. Then the boundary conditions of Neumann problem for this equation in the circle $x^2 + y^2 < 1$ are represented in the form:

$$\frac{\partial^k u(x,y)}{\partial N^k} = f_k(x,y), \quad k = 1, \ldots, \frac{n}{2}, \quad (x,y) \in \Gamma, \tag{6.249}$$

where $f_k(x,y)$ are given real-valued functions. A soluton of equation (6.1) is also searched in the class of real-valued functions. The problem (6.1), (6.249) with $f_k \equiv 0\, (k = 1, \ldots, \frac{n}{2})$ is called homogeneous.

Together with the boundary conditions (6.249) let us also consider the following boundary conditions

$$\frac{\partial^k u(x,y)}{\partial N^k} = g_k(x,y), \quad k = 0, \ldots, \frac{n}{2} - 1, \tag{6.250}$$

where

$$g_0(x,y) = f_1(x,y), \; g_k(x,y) = f_{k+1}(x,y) + k f_k(x,y), \quad k = 1, \ldots, \frac{n}{2} - 1,$$

and $f_k(x,y)$ $(k = 1, \ldots, \frac{n}{2})$ are the functions from (6.249).

According to the assumption, the coefficients of elliptic equation (6.1) are real, hence the problem (6.1), (6.250) is uniquely solvable (see [26]).

The following theorem takes place

Theorem 6.13 *The non-homogeneous problem (6.1), (6.249) is solvable if and only if* $v(0,0) = 0$, *where* $v(x,y)$ *is a solution of the problem (6.1), (6.250). The homogeneous problem (6.1), (6.249) has only one solution* $u(x,y) = const.$

Proof of Theorem 6.13 is similar to that of Theorems 6.11 and 6.12 and is omitted.

Example 6.1. Consider the Bitzadze equation:

$$\frac{\partial^2 u}{\partial \bar{z}^2} = 0, \quad (x,y) \in D, \tag{6.251}$$

where

$$\frac{\partial}{\partial \bar{z}} \equiv \frac{1}{2}\left(\frac{\partial}{\partial x} + i\frac{\partial}{\partial y}\right), \quad z = x + iy, \quad \bar{z} = x - iy.$$

The boundary condition (6.220) for this equation can be written as follows:

$$Re\frac{\partial^k u(x,y)}{\partial N^k} = f_k(x,y), \quad (x,y) \in \Gamma, \quad k = 1,2. \tag{6.252}$$

We are going to find conditions for solvability of the problem (6.251), (6.252) in the explicit form.

The boundary problem (6.221) and (6.222) corresponding to the problem (6.257), (6.252) has the form

$$\frac{\partial^4 \varphi(x,y)}{\partial \overline{z}^2 \partial z^2} = 0, \quad (x,y) \in D, \tag{6.253}$$

$$\frac{\partial^k \varphi(x,y)}{\partial N^k} = g_k(z), \quad (x,y) \in \Gamma, \quad k = 0,1, \tag{6.254}$$

where

$$g_0(z) = f_1(x,y), \quad g_1(z) = f_1(x,y) + f_2(x,y). \tag{6.255}$$

According to Theorem 6.11, the solvability condition for the problem (6.251), (6.252) is

$$\varphi(0,0) = 0 \tag{6.256}$$

where $\varphi(x,y)$ is a solution of the problem (6.253), (6.254).

Let us represent the condition (6.256) by functions $f_1(x,y)$ and $f_2(x,y)$. For this purpose we rewrite a solution of the problem (6.253), (6.254) in the form

$$\varphi(x,y) = Re[\varphi_1(z) + (1 - z\overline{z})\varphi_2(z)], \tag{6.257}$$

where $\varphi_1(z)$ and $\varphi_2(z)$ are analytic in $| z |< 1$ functions satisfying the additional conditions

$$Im\varphi_j(0) = 0, \quad j = 1,2. \tag{6.258}$$

For any analytic functions $\varphi_1(z)$ and $\varphi_2(z)$ the function $\varphi(x,y)$ defined by (6.257) is a solution of equation (6.253). Substituting $\varphi(x,y)$ from (6.257) into (6.254) we get

$$Re\varphi_1(z) = g_0(z), \quad Re[\varphi_1'(z)z - 2\varphi_2(z)] = g_1(z), \quad z \in \Gamma.$$

From these boundary conditions the analytic functions $\varphi_1(z)$ and $\varphi_1'(z)z - 2\varphi_2(z)$ are defined by Schwarz formula:

$$\varphi_1(z) = \frac{1}{2\pi i}\int_\Gamma g_0(\zeta)\frac{\zeta + z}{(\zeta - z)\zeta}d\zeta, \tag{6.259}$$

$$\varphi_1'(z)z - 2\varphi_2(z) = \frac{1}{2\pi i}\int_\Gamma g_1(\zeta)\frac{\zeta + z}{(\zeta - z)\zeta}d\zeta. \tag{6.260}$$

From (6.259) and (6.260) the analytic functions $\varphi_1(z)$ and $\varphi_2(z)$ can be determined uniquely. Substituting these functions into (6.257) we obtain a solution $\varphi(x, y)$ of the problem (6.253), (6.254).

In particular, from (6.257), (6.259) and (6.260), we have

$$\varphi(0,0) = Re[\varphi_1(0) + \varphi_2(0)] = \frac{1}{2\pi i} \int\limits_{\Gamma} [g_0(\zeta) - \frac{1}{2} g_1(\zeta)] ds, \tag{6.261}$$

where s is an arc of the contour Γ.

Substituting $\varphi(0,0)$ from (6.261) into (6.256) and using (6.255) we get

$$\int\limits_{\Gamma} [f_1(x,y) - f_2(x,y)] ds = 0. \tag{6.262}$$

Thus, the solvability condition for the problem (6.251), (6.252) has the form (6.262).

According to Corollary 6.1, the homogeneous problem (6.251), (6.252) (with $f_k = 0$, $k = 1, 2$) has exactly 5 linearly independent solutions.

6.11 Poincaré Problem for Bitzadze Equation

Let D be the unit circle $x^2 + y^2 < 1$ and Γ be the circumference $x^2 + y^2 = 1$. In D we consider Poincaré problem for Bitzadze equation:

$$\frac{\partial^2 u}{\partial \overline{z}^2} = 0, \quad (x, y) \in D, \tag{6.263}$$

$$\frac{\partial u(x,y)}{\partial N} + \alpha u(x,y) + \beta \overline{u(x,y)} = f(x,y), \quad (x,y) \in \Gamma, \tag{6.264}$$

where N is the exterior normal to the boundary Γ at the point $(x, y) \in \Gamma$, and $u(x, y)$ is a solution to be found; α and β are complex constants, $f(x, y)$ is a given function on Γ.

The problem (6.263), (6.264) with $f(x, y) \equiv 0$ is called homogeneous problem. If $\beta = 0$, then it is easy to verify that the functions

$$u(x,y) = z^k \left[1 + (\frac{k}{2} + \frac{\alpha}{2})(1 - z\overline{z})\right], \quad k = 0, 1, \ldots$$

are linearly independent solutions of the homogeneous problem (6.263), (6.264). We assume that $\beta \neq 0$.

The following theorem takes place.

Theorem 6.14 *If $\beta \neq 0$, then the non-homogeneous problem (6.263), (6.264) is always solvable and the corresponding homogeneous problem has exactly four linearly independent solutions.*

Proof. The general solution of equation (6.263) is defined by (see [16])

$$u(x,y) = \varphi_1(z) + \psi_1(z)\overline{z}, \quad (x,y) \in D, \tag{6.265}$$

where $\varphi_1(z)$ and $\psi_1(z)$ are arbitrary analytic in D functions.

We write $\psi_1(z)$ in the form

$$\psi_1(z) = c_0 + z\psi(z), \tag{6.266}$$

where c_0 is a complex constant and $\psi(z)$ is analytic in D function. Substituting $\psi_1(z)$ from (6.266) into (6.265) we get

$$u(x,y) = \varphi(z) + (z\overline{z} - 1)\psi(z) + c_0\overline{z}, \tag{6.267}$$

where

$$\varphi(z) = \varphi_1(z) + \psi(z).$$

Thus, the general solution of equation (6.263) is defined by (6.267), where $\varphi(z)$ and $\psi(z)$ are arbitrary analytic in D functions, and c_0 is an arbitrary complex constant, the functions $\varphi(z)$ and $\psi(z)$ and the constant c_0 being uniquely defined by $u(x,y)$.

Substituting $u(x,y)$ from (6.267) into (6.264) for $|z| = 1$, we get

$$2\psi(z) + \varphi'(z)z + c_0\overline{z} + \alpha\varphi(z) + \alpha c_0\overline{z} + \beta\overline{\varphi(z)} + \beta\overline{c_0}z = f(z) \tag{6.268}$$

where $f(z) = f(x,y)$.

For $|z| = 1$, we have

$$\overline{\varphi(z)} = \overline{\varphi(\frac{1}{\overline{z}})} \quad \text{where} \quad \overline{z} = \frac{1}{z}. \tag{6.269}$$

Taking into account (6.269), the boundary condition (6.268) can be rewritten in the form

$$\phi_1(z) - \phi_2(z) = f(z) \quad \text{for} \quad |z| = 1, \tag{6.270}$$

where

$$\phi_1(z) = 2\psi(z) + z\varphi'(z) + \alpha\varphi(z) + \beta\overline{c}_0 z, \tag{6.271}$$

$$\phi_2(z) = -(\frac{c_0}{z} + \beta\overline{\varphi(\frac{1}{\overline{z}})} + \frac{\alpha c_0}{z}). \tag{6.272}$$

It follows from (6.271) and (6.272) that $\phi_1(z)$ and $\phi_2(z)$ are analytic in the domains $|z| < 1$ and $|z| > 1$, respectively and $\phi_2(z)$ is bounded in the neighborhood of infinity.

Hence, from (6.270) we have (see [15])

$$\phi_1(z) = \phi(z) + c, \quad | z |< 1, \tag{6.273}$$
$$\phi_2(z) = \phi(z) + c, \quad | z |> 1, \tag{6.274}$$

where

$$\phi(z) = \frac{1}{2\pi i}\int_{|\zeta|=1} \frac{f(\zeta)d\zeta}{\zeta - z}, \tag{6.275}$$

$\zeta = \xi + i\eta$ and c is an arbitrary complex constant.

Substituting $\phi_1(z)$ and $\phi_2(z)$ from (6.273) and (6.274) into (6.271) and (6.272) we get

$$2\psi(z) + z\varphi'(z) + \alpha\varphi(z) + \bar{c}_0 z = \phi(z) + c, \quad | z |< 1, \tag{6.276}$$
$$\frac{c_0}{z} + \beta\overline{\varphi(\frac{1}{\bar{z}})} + \frac{\alpha c_0}{z} = -\phi(z) - c, \quad | z |> 1. \tag{6.277}$$

Replacing in (6.277) z by $\frac{1}{\bar{z}}$ we get

$$\bar{c}_0 z + \bar{\beta}\varphi(z) + \overline{\alpha}\bar{c}_0 z = \phi_0(z) - \bar{c}, \quad | z |< 1, \tag{6.278}$$

where

$$\phi_0(z) = \frac{z}{2\pi i}\int_{|\zeta|=1} \frac{f(\zeta)d\zeta}{1 - z\zeta}. \tag{6.279}$$

From (6.278) we have

$$\varphi(z) = \frac{1}{\bar{\beta}}[\phi_0(z) - \bar{c} - \bar{c}_0 z(1 + \overline{\alpha})], \quad | z |< 1. \tag{6.280}$$

Substituting $\varphi(z)$ from (6.280) into (6.276) we get

$$\begin{aligned}\psi(z) \;&=\; \frac{1}{2}\phi(z) - \frac{1}{2\bar{\beta}}[\phi_0'(z)z + \alpha\phi_0(z)] + \frac{c}{2} + \frac{\alpha\bar{c}}{2\beta} \\ &+\; \frac{\bar{c}_0 z}{2\bar{\beta}}(| 1 + \alpha |^2 - | \beta |^2), \quad | z |< 1.\end{aligned} \tag{6.281}$$

Substituting $\varphi(z)$ and $\psi(z)$ from (6.280) and (6.281) into (6.267) we obtain a formula of solution for the problem (6.263), (6.264). This formula includes two arbitrary complex constants c_0 and c_1 or four real constants $Re c_0$, $Im c_0$,

Rec_1, Imc_1. Hence the homogeneous problem (6.263), (6.264) has exactly four linearly independent solutions.

Theorem 6.14 is proved.

Consider now the following boundary conditions for equation (6.263)

$$\frac{\partial u(x,y)}{\partial N} - \alpha(x,y)\overline{u(x,y)} = f(x,y), \quad (x,y) \in \Gamma, \tag{6.282}$$

where $\alpha(x,y)$ and $f(x,y)$ are given functions defined on Γ and $\alpha(x,y) \neq 0$ for $(x,y) \in \Gamma$.

We are going to obtain necessary and sufficient condition in terms of $\alpha(x,y)$ providing unique solvability of the problem (6.263), (6.282) in the circle $|z| < 1$.

Denote by m the index of a function $\alpha(x,y)$ on Γ. By Sokhotzky–Plemelj formula (see [15]), the function $\alpha(x,y)$ can be represented as follows

$$\alpha(x,y) = z^m \frac{\phi^+(z)}{\phi^-(z)}, \quad z = x + iy, \quad |z| = 1, \tag{6.283}$$

where $\phi^+(z)$ and $\phi^-(z)$ are the boundary values on Γ of the function

$$\phi(z) = \exp\left(\frac{1}{2\pi i}\int_\Gamma \frac{ln[\alpha(\frac{\zeta+\bar{\zeta}}{2}, \frac{\zeta-\bar{\zeta}}{2i})\zeta^{-m}]}{\zeta - z} d\zeta\right) \tag{6.284}$$

in the inside and outside of the circle D, respectively.

Taking into account (6.283), the condition (6.282) can be written as follows

$$\frac{1}{\phi^+(z)}\frac{\partial u(x,y)}{\partial N} - \frac{z^m}{\phi^-(z)}\overline{u(x,y)} = f_0(z), \quad (x,y) \in \Gamma, \tag{6.285}$$

where

$$f_0(z) = \frac{f(x,y)}{\phi^+(z)}. \tag{6.286}$$

Substituting the general solution (6.267) of equation (6.263) into (6.285) for $|z| = 1$ we get

$$\frac{1}{\phi^+(z)}\left[z\varphi'(z) + 2\psi(z) + c_0\bar{z}\right] - \frac{z^m}{\phi^-(z)}\left[\overline{\varphi(z)} + \bar{c}_0 z\right] = f_0(z). \tag{6.287}$$

Multiplying both sides of (6.287) by z and using (6.269) for $|z| = 1$ we get

$$\frac{1}{\phi^+(z)}\left[z^2\varphi'(z) + 2\psi(z) + c_0\right] - \frac{z^{m+1}}{\phi^-(z)}\left[\overline{\varphi(\frac{1}{\bar{z}})} + \bar{c}_0 z\right] = zf_0(z). \tag{6.288}$$

Consider the functions $\varphi_1(z)$ and $\psi_1(z)$ defined by

$$\varphi_1(z) = \frac{1}{\phi(z)}\left[z^2\varphi'(z) + 2\psi(z)z + c_0\right], \quad | z |< 1, \tag{6.289}$$

$$\psi_1(z) = \frac{z^{m+1}}{\phi(z)}\left[\overline{\varphi(\frac{1}{\bar{z}})} + \bar{c_0}z\right], \quad | z |> 1. \tag{6.290}$$

In this notation the boundary condition (6.288) takes the form

$$\varphi_1^+(z) - \psi_1^-(z) = zf_0(z), \quad | z |= 1, \tag{6.291}$$

where

$$\varphi_1^+(z) = \lim_{\zeta\to z}\varphi_1(\zeta), \quad \psi_1^-(z) = \lim_{\xi\to z}\psi_1(\xi), \quad | z |= 1, \ | \zeta |< 1, \ | \xi |> 1.$$

It follows from (6.289) and (6.290) that the function $\varphi_1(z)$ is analytic in the circle $| z |< 1$, while the function $\psi_1(z)$ is analytic outside this circle and in the vicinity of infinity satisfies the estimate

$$| \psi_1(z) |\leq c \, | z |^{m+2}. \tag{6.292}$$

Thus, from (6.291) we get (see [15])

$$\varphi_1(z) = \omega(z) + P_{m+2}(z), \quad | z |< 1, \tag{6.293}$$

$$\psi_1(z) = \omega(z) + P_{m+2}(z), \quad | z |> 1, \tag{6.294}$$

where

$$\omega(z) = \frac{1}{2\pi i}\int\limits_{\Gamma}\frac{\zeta f_0(\zeta)d\zeta}{\zeta - z}, \quad | z |\neq 1, \tag{6.295}$$

and $P_{m+2}(z)$ is an arbitrary polynomial of degree at most $m+2$ and $P_{m+2}(z) \equiv 0$ for $m + 2 < 0$.

Substituting $\varphi_1(z)$ and $\psi_1(z)$ from (6.293) and (6.294) into (6.289) and (6.290) we get the following equations

$$\frac{1}{\phi(z)}\left[z^2\varphi'(z) + z\psi(z) + c_0\right] = \omega(z) + P_{m+2}(z), \quad | z |< 1, \tag{6.296}$$

$$\frac{z^{m+1}}{\phi(z)}\left[\overline{\varphi(\frac{1}{\bar{z}})} + \bar{c_0}z\right] = \omega(z) + P_{m+2}(z), \quad | z |> 1 \tag{6.297}$$

to determine $\varphi(z)$ and $\psi(z)$.

Consider the following cases.

Case I. Let $m = -2$. Then $P_{m+2}(z) = c = const$ and equations (6.296) and (6.297) take the form

$$\frac{1}{\phi(z)}\left[z^2\varphi'(z) + 2\psi(z)z + c_0\right] = \omega(z) + c, \quad | z |< 1, \tag{6.298}$$

$$\frac{1}{z\phi(z)}(\overline{\varphi(\frac{1}{\bar{z}})} + \bar{c}_0 z) = \omega(z) + c, \quad | z |> 1. \tag{6.299}$$

Substituting in (6.298) $z = 0$ we get

$$c = \frac{c_0}{\phi(0)} - \omega(0). \tag{6.300}$$

Substituting c from (6.300) into (6.298) and (6.299) we obtain

$$\psi(z) = \frac{\omega(z) - \omega(0)}{z}\phi(z) - z\varphi'(z) + \frac{(\phi(z) - \phi(0))c_0}{\phi(0)\phi(z)}, \quad | z |< 1, \tag{6.301}$$

$$\frac{1}{z\phi(z)}\left[\overline{\varphi(\frac{1}{\bar{z}})} + \bar{c}_0 z\right] = \omega(z) + \frac{c_0}{\phi(0)} - \omega(0), \quad | z |> 1. \tag{6.302}$$

Passing in (6.302) to the limit as $| z |\to +\infty$, we get

$$\bar{c}_0 = \frac{c_0}{\phi(0)} - \omega(0). \tag{6.303}$$

Taking into account (6.303), equation (6.302) can be written as follows

$$\frac{1}{z\phi(z)}\left[\overline{\varphi(\frac{1}{\bar{z}})} + \bar{c}_0 z\right] = \omega(z) + \bar{c}_0, \quad | z |> 1. \tag{6.304}$$

Replacing in (6.304) z by $\frac{1}{\bar{z}}$ we get

$$\varphi(z) = \frac{1}{z}\overline{\omega(\frac{1}{\bar{z}})\phi(\frac{1}{\bar{z}})} + \left[\overline{\phi(\frac{1}{\bar{z}})} - 1\right]\frac{c_0}{z}, \quad | z |< 1. \tag{6.305}$$

Since $\omega(\infty) = 0$ and $\phi(\infty) = 1$, for any complex constant c_0 the function $\varphi(z)$ defined by (6.305) is analytic in the circle $| z |< 1$.

Equation (6.303) is uniquely solvable with respect to c_0 if and only if

$$| \phi(0) |\neq 1. \tag{6.306}$$

It follows from (6.284) that the condition (6.306) can be written as

$$\int_0^{2\pi} ln \mid \alpha(\cos\theta, \sin\theta) \mid d\theta \neq 0. \tag{6.307}$$

Thus, if $m = -2$ the problem (6.263), (6.282) is uniquely solvable if and only if the condition (6.307) is satisfied.

Let

$$\int_{0}^{2\pi} ln \mid \alpha(\cos\theta, \sin\theta) \mid d\theta = 0. \tag{6.308}$$

In this case we have $\mid \phi(0) \mid= 1$, and hence equation (6.303) takes the form

$$c_0\bar{\delta} - \bar{c_0}\delta = \omega(0)\delta, \tag{6.309}$$

where $\delta = \sqrt{\phi(0)}$.

From (6.309) we have

$$Re[\omega(0)\delta] = 0. \tag{6.310}$$

Let the condition (6.310) be satisfied. Then the general solution of equation (6.309) is defined by

$$c_0 = \frac{1}{2}\omega(0)\delta^2 + c_1\delta,$$

where c_1 is an arbitrary real constant.

Taking into account (6.295), the condition (6.310) can be represented as

$$Re\left[\frac{\delta}{2\pi i}\int_{|z|=1}\frac{f(z)}{\phi^+(z)}dz\right] = 0,$$

where $f(z) = f(x, y)$.

Thus we have proved the following

Theorem 6.15 *For $m = -2$ the problem (6.263), (6.282) is uniquely solvable if and only if the function $\alpha(x, y)$ satisfies the condition (6.307). If this condition is satisfied, then a solution of this problem is defined by (6.267) where $\varphi(z)$, $\psi(z)$ and c_0 are uniquely determined by (6.301), (6.303) and (6.305).*

Case II. Let $m + 2 \neq 0$. Using equations (6.296), (6.297) the problem (6.263), (6.282) can be investigated in a similar way. In this case it can be proved that either the homogeneous problem has non-trivial solution (for $m + 2 > 0$) or the non-homogeneous problem is solvable for any smooth functions $f(x, y)$ (for $m + 2 > 0$). These results can be sumarized as:

Theorem 6.16 *For $m + 2 > 0$ the non-homogeneous problem (6.263), (6.282) is always solvable and the corresponding homogeneous problem has exactly $2(m + 2)$ linearly independent solutions.*

Theorem 6.17 *For $m+2<0$ the homogeneous problem (6.263), (6.282) has only zero solution and the corresponding non-homogeneous problem is solvable if and only if the function $f(x,y)$ satisfies $-2(m+2)$ conditions of the form*

$$Re\left[\int_\Gamma f(x,y)\psi_j(x,y)ds\right]=0, \quad (j=0,1,\ldots,-2(m+2)),$$

where $\psi_1(x,y),\ldots,\psi_k(x,y)$ $(k=-2(m+2))$ are some linearly independent functions.

From Theorems 6.15 – 6.17 follow

Theorem 6.18 *The problem (6.263), (6.282) is uniquely solvable if and only if $m=-2$ and the condition (6.307) is satisfied.*

Now let us consider the following boundary condition

$$\frac{\partial u(x,y)}{\partial N}+\alpha u(x,y)-\beta z^{-2}\overline{u(x,y)}=f(x,y), \quad (x,y)\in\Gamma, \tag{6.311}$$

where α and β are complex constants.

The problem (6.263), (6.311) can be investigated as the problem (6.263), (6.264). We formulate the corresponding result without proof.

Theorem 6.19 *The problem (6.263), (6.311) is uniquely solvable if and only if α and β satisfy the conditions*

$$\beta\neq 0, \quad |\,\beta\,|\neq|\,1+\alpha\,|\,. \tag{6.312}$$

The obtained results show that for unique solvability of Poincaré problem for non-regular elliptic equations the presence of $\overline{u(x,y)}$ in the boundary condition is necessary.

Chapter 7

Some Classes of Improperly Elliptic Equations

Chapter 7 is a supplement to chapter 6. The results of this chapter have been used in chapter 6 to investigate the boundary value problems for improperly elliptic equations.

7.1 Improperly Elliptic Equations in a Class of Real-Valued Functions

Let D be a simple connected domain in the plane. In D we consider the equation

$$\sum_{k=0}^{n} A_k \frac{\partial^n u}{\partial x^k \partial y^{n-k}} = 0, \tag{7.1}$$

where $A_0, A_1, \ldots, A_n$ are complex constants and $A_0 \neq 0$.

Let $\lambda_1, \ldots, \lambda_n$ be the roots of the characteristic equation

$$A_0 \lambda^n + A_1 \lambda^{n-1} + \cdots + A_n = 0. \tag{7.2}$$

In this section we assume that equation (7.1) is elliptic and

$$Im\lambda_k > 0, \quad k = 1, 2, \ldots, n. \tag{7.3}$$

One has the following

Theorem 7.1 *If the condition (7.3) is satisfied, then the number of linearly independent real-valued solutions of equation (7.1) is equal to* n^2.

In order to prove Theorem 7.1 we are going to prove some preliminary lemmas. Consider the set of functions

$$y^l(x+\mu_j y)^{m-l} \quad (l = 0,\dots,k_j-1; \quad j = 1,\dots,p), \tag{7.4}$$

where m, p, $k_1,\dots,k_p$ are some positive integers and $\mu_1,\dots,\mu_p$ are complex constants satisfying

$$\mu_j \neq \mu_l \quad for \quad j \neq l; \quad j,l = 1,\dots,p.$$

Lemma 7.1 *If* $k_1+k_2+\cdots+k_p \leq m+1$, *then the set of functions (7.4) is linearly independent (over the field of complex numbers).*

Proof. For simplicity, we prove this lemma in the case, where $p = k_1 = k_2 = 2$, and $m \geq 3$ (the general case can be treated in a similar way). In the mentioned case the set of functions (7.4) takes the form

$$(x+\mu_1 y)^m, y(x+\mu_1 y)^{m-1}, (x+\mu_2 y)^m, y(x+\mu_2 y)^{m-1}. \tag{7.5}$$

Let

$$c_1(x+\mu_1 y)^m + c_2 y(x+\mu_1 y)^{m-1} + c_3(x+\mu_2 y)^m + c_4 y(x+\mu_2 y)^{m-1} \equiv 0, \tag{7.6}$$

where c_1, c_2, c_3 and c_4 are complex constants. We are going to show that (7.6) implies $c_j = 0$ $(j = 1,\dots,4)$.

Applying the operator

$$(\frac{\partial}{\partial x} - \mu_1 \frac{\partial}{\partial y})^2 (\frac{\partial}{\partial x} - \mu_2 \frac{\partial}{\partial y})$$

to both sides of (7.5) we get $a_0 c_4 (x+\mu_2 y)^{m-3} \equiv 0$, where a_0 is a non-zero constant. Hence $c_4 = 0$. Applying to both sides of (7.6) the operator $(\frac{\partial}{\partial x} - \mu_1 \frac{\partial}{\partial y})^2$, we get $c_3 = 0$, because $c_4 = 0$. Proceeding in such a way we get $c_j = 0$ $(j = 1,2,3,4)$. Hence for $m \geq 3$ the set of functions (7.5) is linearly independent.

Lemma 7.1 is proved.

Let

$$k_1+k_2+\cdots+k_p = m+1.$$

The set (7.4) includes $m+1$ functions, hence according to lemma 7.1 these functions are linearly independent.

Denoting these functions by $u_1(x,y),\ldots,u_{m+1}(x,y)$, it is easy to see that these functions admit the representation

$$u_k(x,y)=\sum_{j=0}^{m}a_{kj}x^k y^{m-k},\quad k=1,\ldots,m+1,$$

where a_{kj} are some complex constants.

Let Δ_m be the matrix, whose columns are the vectors $(a_{10},a_{20},\ldots,a_{m+1,0})$, $(a_{11},a_{21},\ldots,a_{m+1,1})$, $\ldots$, $(a_{1m},a_{2m},\ldots,a_{m+1,m})$.

Lemma 7.2 *The determinant of matrix Δ_m is not equal to zero.*

Proof is obvious.

Consider the following equation

$$Im(Az)\;=\;0, \tag{7.7}$$

where A is a square matrix of order n and $z=(z_1,\ldots,z_n)$ is unknown vector-column. Here the elements of matrix A and vector z are complex numbers.

Lemma 7.3 *The number k_0 of linearly independent solutions of equation (7.7) (over the field of real numbers) is defined by*

$$k_0=2n-rang(A,\bar{A}). \tag{7.8}$$

Proof. Changing in (7.7) the variable $z=x+iy$ where $x=Rez$ and $y=Imz$, we get

$$ImAx+ReAy=0. \tag{7.9}$$

The numbers of linearly independent solutions of the systems (7.7) and (7.9) are the same. It is known (see [30]), that the number of linearly independent solutions of the system (7.9) is defined by

$$k_0=2n-rang(ImA,ReA). \tag{7.10}$$

We have

$$rang(A,\bar{A})=rang(A-\bar{A},A+\bar{A})=rang(ImA,ReA). \tag{7.11}$$

From (7.10) and (7.11) we obtain (7.8).

Lemma 7.3 is proved.

Proof of Theorem 7.1. Without loss of generality we assume that $(0,0)\in D$. Let the roots $\lambda_1,\ldots,\lambda_n$ of the characteristic equation (7.2) be simple. Then the general solution of equation (7.1) in a simple connected domain D is defined by (see [16])

$$u(x,y) = \sum_{j=1}^{n} \varphi_j(x + \lambda_j y), \tag{7.12}$$

where $\varphi_1(x + \lambda_1 y), \ldots, \varphi_n(x + \lambda_n y)$ are arbitrary analytic functions on $x + \lambda_1 y, \ldots, x + \lambda_n y$, respectively $((x,y) \in D)$.

Consider the Taylor series expansion of the function $\varphi_j(x+\lambda_j y)$ in a vicinity of the point $(0,0) \in D$,

$$\varphi_j(x + \lambda_j y) = \sum_{l=0}^{\infty} c_{lj}(x + \lambda_j y)^l \quad for \quad x^2 + y^2 \leq \varepsilon^2, \tag{7.13}$$

where ε is a sufficiently small positive number.

Substituting $\varphi_j(x + \lambda_j y)$ from (7.13) into (7.12) we get

$$u(x,y) = \sum_{l=0}^{\infty} \sum_{j=1}^{n} c_{lj}(x + \lambda_j y)^l, \quad x^2 + y^2 \leq \varepsilon^2. \tag{7.14}$$

Now let $u(x,y)$ be a real-valued, i.e.

$$u(x,y) - \overline{u(x,y)} \equiv 0, \quad (x,y) \in D. \tag{7.15}$$

Substituting $u(x,y)$ from (7.14) into (7.15) we get

$$\sum_{j=1}^{n} [c_{lj}(x + \lambda_j y)^l - \bar{c}_{lj}(x + \bar{\lambda}_j y)^l] \equiv 0, \quad (l = 0, 1, \ldots). \tag{7.16}$$

From (7.3) and lemma 7.1 it follows that for any positive integer $k \geq 2n - 1$ the set of functions

$$(x + \lambda_j y)^k, \quad (x + \bar{\lambda}_j y)^k, \quad j = 1, \ldots, n$$

is linearly independent (over the field of complex numbers). So from (7.16) we have

$$c_{lj} = 0, \quad j = 1, \ldots, n, \quad l \geq 2n - 1. \tag{7.17}$$

Taking into account (7.17) the formula (7.14) can be written in the form

$$u(x,y) = \sum_{l+j \leq n-1} a_{lj} x^l y^j + \sum_{l=n}^{2n-2} \sum_{j=1}^{n} c_{lj}(x + \lambda_j y)^l, \quad x^2 + y^2 \leq \varepsilon^2, \tag{7.18}$$

where a_{lj} are arbitrary real constants and $c_{lj}, \ j = 1, \ldots, n, \ l = n, \ldots, 2n-2$ satisfy the relations (7.16).

We have

$$(x + \lambda_j y)^l \equiv \sum_{k=0}^{l} b_{lkj} x^k y^{l-k}, \tag{7.19}$$

$$(x + \bar{\lambda}_j y)^l \equiv \sum_{k=0}^{l} \bar{b}_{lkj} x^k y^{l-k}, \tag{7.20}$$

where b_{lkj} are some constants.

Substituting $(x+\lambda_j y)^l$ and $(x+\overline{\lambda_j y})^l$ from (7.19) and (7.20) into (7.16) we get

$$\sum_{j=1}^{n} [b_{lkj} c_{lj} - \bar{b}_{lkj} \bar{c}_{lj}] = 0, \quad k = 0, \ldots, l, \quad (l = n, n+1, \ldots, 2n-1). \tag{7.21}$$

Let b_{lj} denote the following vector-column

$$b_{lj} = (b_{l0j}, b_{l1j}, \ldots, b_{llj}),$$

and let B_l denote the matrix the rows of which are the vectors $b_{l1}, b_{l2}, \ldots, b_{ln}$. Let l be a positive integer satisfying $n \leq l \leq 2n-2$.

According to Lemma 7.1, the set of functions

$$(x + \lambda_j y)^l, (x + \bar{\lambda}_k y)^l, \quad j = 1, \ldots, n; \quad k = 1, \ldots, l+1-n$$

is linearly independent. By Lemma 7.2 the set of vectors

$$b_{lj}, \ \bar{b}_{lk}, \quad j = 1, \ldots, n; \quad k = 1, \ldots, l+1-n,$$

is also linearly independent (over the field of complex numbers). Hence

$$rang(B_l, \overline{B_l}) = l + 1. \tag{7.22}$$

The system (7.21) can be written as follows

$$Im[B_l c_l] = 0, \tag{7.23}$$

where $c_l = (c_{l1}, \ldots, c_{ln})$ is a vector-column to be found. According to Lemma 7.3 and (7.22), the system (7.23) has $2n - l - 1$ linearly independent solutions. Let

$$\theta_l = 2n - l - 1, \tag{7.24}$$

and let $\alpha_{1l}, \ldots, \alpha_{\theta_l l}$ be the linearly independent solutions of (7.23). Then the general solution of (7.23) is defined by

$$c_l = d_{l1} \alpha_{1l} + \cdots + d_{l\theta_l} \alpha_{\theta_l l}. \tag{7.25}$$

Recall that the components of vector c_l are $c_{l1}, \ldots, c_{ln}$. Substituting c_l from (7.25) into (7.18) we get

$$u(x,y) = \sum_{l+j\leq n-1} a_{lj}x^l y^j + \sum_{l=n}^{2n-2}\sum_{\rho=1}^{\theta_l} d_{l\rho}u_{l\rho}(x,y), \quad x^2+y^2 \leq \varepsilon^2, \tag{7.26}$$

where

$$u_{l\rho}(x,y) = \sum_{j=1}^{n} \alpha_{\rho lj}(x+\lambda_j y)^l,$$

and $\alpha_{\rho l1}, \ldots, \alpha_{\rho ln}$ are the components of vector $\alpha_{\rho l}$ ($\rho = 1, \ldots, \theta_l$). Since the vector $\alpha_{\rho l}$ is a solution of the system (7.23) the functions $u_{l\rho}(x,y)$ are real-valued.

According to Lemma 7.1, for any fixed $l \geq n-1$ the functions $(x+\lambda_1 y)^l, \ldots, (x+\lambda_n y)^l$ are linearly independent. Hence from the linear independence of vectors $\alpha_{1l}, \ldots, \alpha_{\theta_l l}$, it follows that for any positive integer l ($n \leq l \leq 2n-2$), the set of vectors $u_{l1}(x,y), \ldots, u_{l\theta_l}(x,y)$ is linearly independent. Since $u_{lj}(x,y)$ ($j = 1, \ldots, \theta_l$) are homogeneous polynomials of degree l with respect to x and y, the set of vectors

$$x^l y^j,\ u_{k\rho}(x,y),\ 0 \leq l+j \leq n-1,\ \rho = 1, \ldots, \theta_k,\ k = n, \ldots, 2n-2 \tag{7.27}$$

is linearly independent, too. It is easy to check that (7.27) contains n^2 functions. Since both sides of (7.26) satisfy (7.1) and coincide in a vicinity of the point $(0,0)$, they are equal in the domain D, i.e.

$$u(x,y) = \sum_{l+j\leq n-1} a_{lj}x^l y^j + \sum_{l=n}^{2n-2}\sum_{\rho=1}^{\theta_l} d_{l\rho}u_{l\rho}(x,y), \quad (x,y) \in D, \tag{7.28}$$

where a_{lj} and $d_{l\rho}$ are arbitrary real constants. From (7.28) follows the assertion of Theorem 7.1 in the case where all the roots of the characteristic equation are simple.

Now let $\lambda_1, \ldots, \lambda_n$ be distinct roots of the characteristic equation (7.2) and let $k_1, \ldots, k_m$ be their multiplicities ($k_1 + \cdots + k_m = n$). Then the general solution of equation (7.1) in a simple connected domain D is defined by (see [11])

$$u(x,y) = \sum_{j=1}^{m}\sum_{l=0}^{k_j-1} y^l \varphi_{jl}(x+\lambda_j y), \tag{7.29}$$

where $\varphi_{jl}(x+\lambda_j y)$ ($l = 0, \ldots, k_j - 1$) is an arbitrary analytic on $x+\lambda_j y$ function at $(x,y) \in D$.

Using (7.29) and the above arguments, we complete our proof of Theorem 7.1.

Now we describe a simple method of constructing the real-valued solutions of equation (7.1) (without using the roots of characteristic equation (7.2)).

Let

$$A_k = a_k + ib_k,$$

where a_k and b_k are real. Equation (7.1) in the class of real-valued functions can be represented as a system of equations

$$\sum_{k=0}^{n} a_k \frac{\partial^n u}{\partial x^k \partial y^{n-k}} = 0, \tag{7.30}$$

$$\sum_{k=0}^{n} b_k \frac{\partial^n u}{\partial x^k \partial y^{n-k}} = 0, \tag{7.31}$$

where $a_k = ReA_k$ and $b_k = ImA_k$.

A solution of the system (7.30), (7.31) is searched as

$$u(x, y) = \sum_{k=0}^{l} c_k x^k y^{l-k}, \tag{7.32}$$

where l is a fixed positive integer satisfying $n \leq l \leq 2n - 2$ and $c_0, \ldots, c_l$ are real constants to be found.

Substituting $u(x, y)$ from (7.32) into (7.30) and (7.31) and equalizing the coefficients of the corresponding powers of $x^k y^{l-n-k}$ $(k = 0, \ldots, l - n)$ we obtain the system of equations

$$B_{k0}c_0 + B_{k1}c_1 + \cdots + B_{kl}c_l = 0, \quad k = 1, \ldots, 2(l - n + 1), \tag{7.33}$$

where B_{kj} are well defined real constants.

Let r_l be the rank of principal matrix of the system (7.33). Since $n \leq l \leq 2n - 2$ we have $2(l - n + 1) < l + 1$. Hence

$$r_l \leq 2(l - n + 1). \tag{7.34}$$

We set

$$\beta_l = 2(l - n + 1) - r_l, \tag{7.35}$$

From (7.34) we have

$$\beta_l \geq 0. \tag{7.36}$$

It is known that the number δ_l of linearly independent solutions of the system (7.33) is defined by

$$\delta_l = l + 1 - r_l.$$

From here and (7.35) we have

$$\delta_l = 2n - l - 1 + \beta_l. \tag{7.37}$$

Substituting the linearly independent solutions of (7.33) into (7.32), we obtain δ_l linearly independent solutions of the system (7.30), (7.31). Denote these solutions by $u_{l\rho}(x,y)$ $(\rho = 1, \ldots, \delta_l)$.

It is easy to check that the functions

$$x^k y^j,\ u_{l\rho}(x,y),\ 0 \le k+j \le n-1,\ \rho = 1, \ldots, \delta_l,\ l = n, \ldots, 2n-2 \tag{7.38}$$

are linearly independent real-valued solutions of the system (7.30), (7.31).

It follows from (7.37) that the number of functions in (7.38) is equal to

$$n^2 + \beta_n + \beta_{n+1} + \cdots + \beta_{2n-2}.$$

Let the condition (7.3) be satisfied. Then the number of linearly independent solutions of the system (7.30), (7.31) is equal to n^2 (see Theorem 7.1). Hence

$$n^2 + \beta_n + \beta_{n+1} + \cdots + \beta_{2n-2} \le n^2.$$

From here and (7.36) we obtain

$$\beta_k = 0, \quad k = n, n+1, \ldots, 2n-2. \tag{7.39}$$

Hence the set (7.38) consists of exactly n^2 functions and these functions form a fundamental system of solutions of (7.1) in the class of real-valued functions. From (7.35), (7.37) and (7.39) we have

$$r_l = 2(l - n + 1), \quad \delta_l = 2n - l - 1.$$

Thus, the principal matrix of the system (7.33) posesses a non-zero minor of order $2(l-n-1)$. This fact essentially simplifies the process of determining linearly independent solutions of system (7.33).

7.2 Two Elliptic Equations with One Unknown Function

Let D be a simple connected domain and $(0,0) \in D$. In this domain we consider the following system of two improperly elliptic equations

$$\sum_{k=0}^{n} A_k \frac{\partial^n u(x,y)}{\partial x^k \partial y^{n-k}} = 0, \quad (x,y) \in D, \tag{7.40}$$

$$\sum_{k=0}^{m} B_k \frac{\partial^m u(x,y)}{\partial x^k \partial y^{n-k}} = 0, \quad (x,y) \in D, \tag{7.41}$$

where A_k and B_k are complex constants and $A_0 \neq 0$, $B_0 \neq 0$.

Let

$$A_0\lambda^n + A_1\lambda^{n-1} + \cdots + A_n = 0, \tag{7.42}$$

and

$$B_0\mu^m + B_1\mu^{m-1} + \cdots + B_m = 0 \tag{7.43}$$

be the characteristic equations of (7.40) and (7.41), respectively.

We assume that $n \geq m \geq 1$ and

$$Im\lambda_k > 0, \quad k = 1, \ldots, n, \tag{7.44}$$

$$Im\mu_k < 0, \quad k = 1, \ldots, m, \tag{7.45}$$

where $\lambda_1, \ldots, \lambda_n$ and $\mu_1, \ldots, \mu_m$ are the roots of (7.42) and (7.43), respectively.

One has the following

Theorem 7.2 *Let the conditions (7.44) and (7.45) be satisfied. Then the system of equations (7.40), (7.41) has exactly nm linearly independent solutions over the field of complex numbers.*

Proof. For simplicity, we assume that the roots $\lambda_1, \ldots, \lambda_n$ and $\mu_1, \ldots, \mu_m$ of equations (7.42) and (7.43) are simple and $2 \leq m \leq n$. Then the general solution of equation (7.41) is defined by (see [11]),

$$u(x, y) = \sum_{k=1}^{m} \varphi_k(x + \mu_k y), \tag{7.46}$$

where $\varphi_k(x + \mu_k x)$ is an arbitrary analytic function on $x + \mu_k y$ ($(x, y) \in D$, $k = 1, \ldots, m$).

Consider the Taylor series expansion of a function φ_k in a vicinity of the point $(0, 0) \in D$,

$$\varphi_k(x + \mu_k y) = \sum_{j=0}^{\infty} c_{kj}(x + \mu_k y)^j, \tag{7.47}$$

where c_{kj} are some constants.

Substituting $\varphi_k(x + \mu_k y)$ from (7.47) into (7.46) we get

$$u(x, y) = \sum_{j=0}^{\infty} u_j(x, y) \quad for \quad | x + iy | < \varepsilon, \tag{7.48}$$

where ε is a sufficiently small positive number and

$$u_j(x, y) = \sum_{k=1}^{m} c_{kj}(x + \mu_k y)^j. \tag{7.49}$$

Let $u(x, y)$ be a solution of equation (7.40). Using the above arguments we conclude that $u(x, y)$ admits Taylor series expansion in a vicinity of the point $(0, 0)$.

$$u(x, y) = \sum_{j=0}^{\infty} v_j(x, y) \quad for \quad | x + iy |< \varepsilon, \tag{7.50}$$

where

$$v_j(x, y) = \sum_{l=1}^{n} d_{lj}(x + \lambda_l y)^j,$$

and d_{lj} are some constants.

Now let $u(x, y)$ be a solution of the system (7.40), (7.41). Then this solution admits simultaneously two series expansions (7.48) and (7.50). Comparing these two series we get

$$\sum_{k=1}^{m} c_{kj}(x + \mu_k y)^j \equiv \sum_{l=1}^{n} d_{lj}(x + \lambda_l y)^j, \quad | x + iy |< \varepsilon. \tag{7.51}$$

Let $j \geq m+n-1$. Then from (7.44), (7.45) and Lemma 7.1 it follows that the functions

$$(x + \mu_k y)^j, \quad (x + \lambda_l y)^j, \quad k = 1, \ldots, m, \quad l = 1, \ldots, n$$

are linearly independent. Hence from (7.51) we have

$$c_{kj} = 0, \quad d_{lj} = 0, \quad j \geq m + n - 1; \quad k = 1, \ldots, m; \quad l = 1, \ldots, n. \tag{7.52}$$

The series (7.48) can be rewritten as follows

$$u(x, y) = \sum_{j=0}^{m+n-2} u_j(x, y), \quad | x + iy |< \varepsilon. \tag{7.53}$$

Since both sides of (7.53) are analytic in D on real variables x and y, and coincide in a vicinity of the point $(0, 0)$, they are equal in D, i.e.

$$u(x, y) = \sum_{j=0}^{m+n-2} u_j(x, y), \quad (x, y) \in D. \tag{7.54}$$

Now let the index j in (7.51) be fixed and satisfy the inequality

$$n \leq j \leq n + m - 2.$$

Then, according to Lemma 7.1, the functions

$$\{(x+\lambda_l y)^j, \quad (x+\mu_k y)^j, \quad l=1,\dots,n; \quad k=1,\dots,j+1-n\} \tag{7.55}$$

are linearly independent. These functions are homogeneous polynomials of degree j on x and y and their number is equal to $j+1$. On the other hand, the number of linearly independent homogeneous polynomials of degree j (on x and y) is also equal to $j+1$.

Thus, for $p=j+2-n,\dots,m$ the functions $(x+\mu_p y)^j$ can be expressed linearly by the functions (7.55), i.e.

$$(x+\mu_p y)^j = \sum_{l=1}^{n} a_{jpl}(x+\lambda_l y)^j + \sum_{k=1}^{j+1-n} b_{jpk}(x+\mu_k y)^j, \tag{7.56}$$

where a_{jpl} and b_{jpk} are well defined constants.

Substituting $(x+\mu_p y)^j$ $(p=j+2-n,\dots,m)$ from (7.56) into (7.51) and taking into account that the set of functions (7.55) is linearly independent, we get

$$c_{kj} = -\sum_{p=j+2-n}^{m} c_{pj} b_{jpk}, \quad k=1,\dots,j+1-n, \tag{7.57}$$

$$d_{kj} = \sum_{p=j+2-n}^{m} a_{jpk} c_{pj}, \quad k=1,\dots,n. \tag{7.58}$$

Thus, for $n \le j \le n+m-2$ the constants $c_{1j},\dots,c_{mj}$ and $d_{1j},\dots,d_{nj}$ satisfy the identity (7.51) if and only if they satisfy the relations (7.57) and (7.58). Substituting c_{kj} $(k=1,\dots,j+1-n)$ from (7.57) into (7.49) and (7.51) we get

$$u_j(x,y) = \sum_{k=j+2-n}^{m} c_{kj} w_{jk}(x,y), \tag{7.59}$$

$$\sum_{k=j+2-n}^{m} c_{kj} w_{jk}(x,y) \equiv \sum_{l=1}^{n} d_{lj}(x+\lambda_l y)^j, \tag{7.60}$$

where

$$w_{jk}(x,y) = (x+\mu_k y)^j - \sum_{l=1}^{j+1-n} b_{jkl}(x+\mu_l y)^j, \; k=j+2-n,\dots,m, \tag{7.61}$$

the constants d_{kj} are defined by (7.58), while c_{jk} are arbitrary complex constants, and

$$n \le j \le n+m-2. \tag{7.62}$$

It is clear that the function $w_{jk}(x,y)$ is homogeneous polynomial of degree j on real variables x and y.

By (7.49) and (7.59) the function (7.54) can be represented as follows

$$u(x,y) = \sum_{j+k\leq m-1} \beta_{kj} x^k y^j + \sum_{j=m}^{n-1} \sum_{k=1}^{m} c_{kj}(x+\mu_k y)^j$$
$$+ \sum_{j=n}^{m+n-2} \sum_{p=j+2-n}^{m} c_{pj} w_{jp}(x,y), \quad (x,y) \in D. \tag{7.63}$$

Since $j \geq m$, then according to Lemma 7.1 the functions $(x+\mu_k y)^j$ $(k=1,\ldots,m)$ are linearly independent. Hence the functions $w_{jp}(x,y)$, $(j = n,\ldots,m+n-2;\ p = j+2-n,\ldots,m)$ are also linearly independent, and in view of (7.60), (7.61) and (7.62) they satisfy the system (7.40), (7.41). On the other hand, it is clear that the functions $x^k y^j$ $(k+j \leq m-1)$ and $(x+\mu_k y)^j$ $(k=1,\ldots,m;\ j=m,\ldots,n-1)$ also satisfy this system.
Hence for any complex constants β_{kj}, c_{kj} and c_{pj} the function $u(x,y)$ defined by (7.63) is a solution of the system (7.40), (7.41).

Therefore the set of functions

$$x^j y^k,\ (j+k \leq n-1);\quad (x+\mu_k y)^j,\ (k=1,\ldots,m;\ j=m,\ldots,n-1),$$
$$w_{jk}(x,y),\quad (k=j+2-n,\ldots,m;\ j=n,\ldots,m+n-2)$$

forms a complete set of linearly independent solution of the system (7.40), (7.41) and their number is equal to mn.

Theorem 7.2 is proved.

From Theorem 7.2 we obtain the following

Corollary 7.1 *If the conditions (7.44), (7.45) are satisfied, then the system of equations (7.40), (7.41) has exactly* $2mn$ *linearly independent solutions over the field of real numbers.*

Indeed, if $u_1(x,y),\ldots,u_l(x,y)$ $(l = mn)$ are the linearly independent solutions of the system (7.40), (7.41) over the field of complex numbers, then the functions $u_1(x,y),\ldots,u_l(x,y)$, $iu_1(x,y),\ldots,iu_l(x,y)$ form a complete set of linearly independent solutions of this system over the field of real numbers.

Consider now the operators

$$L_p u \equiv \sum_{k=0}^{p} A_k \frac{\partial^p u(x,y)}{\partial x^k \partial y^{p-k}}, \tag{7.64}$$

$$L_q u \equiv \sum_{k=0}^{q} B_k \frac{\partial^q u(x,y)}{\partial x^k \partial y^{q-k}}, \tag{7.65}$$

$$L_q^* u \equiv \sum_{k=0}^{q} \overline{B_k} \frac{\partial^q u(x,y)}{\partial x_k \partial y^{q-k}}, \tag{7.66}$$

where A_k and B_k are complex constants and $A_0 \neq 0,\ B_0 \neq 0$.

Let

$$A_0 \lambda^p + A_1 \lambda^{p-1} + \ldots + A_p = 0, \tag{7.67}$$

and

$$B_0 \mu^q + B_1 \mu^{q-1} + \ldots + B_q = 0, \tag{7.68}$$

be the characteristic equations of $L_p u = 0$ and $L_q u = 0$, respectively, and let $\lambda_1, \ldots, \lambda_p$ and $\mu_1, \ldots, \mu_q$ be the roots of these equations.

We assume that

$$Im\lambda_k > 0, \quad k = 1, \ldots, p, \tag{7.69}$$

$$Im\mu_k > 0, \quad k = 1, \ldots, q. \tag{7.70}$$

In the simple connected domain D consider the system of equations

$$L_p L_q u = 0, \tag{7.71}$$

$$L_q L_q^* u = 0. \tag{7.72}$$

One has the following

Theorem 7.3 *If the conditions (7.69), (7.70) are satisfied, then the number of linearly independent real-valued solutions of the system (7.71), (7.72) is equal to* $2pq + q^2$.

Proof. Let $v(x,y)$ be a solution of the equation

$$L_q^* v = 0, \quad (x,y) \in D. \tag{7.73}$$

Then the function

$$u(x,y) = Rev(x,y) \tag{7.74}$$

is a real-valued solution of equation (7.72). The converse is also true, if $u(x,y)$ is a real-valued solution of equation (7.72) it can be represented as in (7.74). This follows from formula (7.29) with $m = q$ and $n = 2q$.

The formula (7.74) can be rewritten as follows

$$u(x,y) = \frac{1}{2}(v(x,y) + \overline{v(x,y)}). \tag{7.75}$$

From (7.73) we get

$$L_q \overline{v(x,y)} = 0, \quad (x,y) \in D. \tag{7.76}$$

Substituting $u(x,y)$ from (7.75) into (7.71) and taking into account (7.76) we obtain

$$L_p L_q v(x,y) = 0, \quad (x,y) \in D. \tag{7.77}$$

It is clear that the roots of the equation

$$\overline{B_0}\mu^q + \overline{B_1}\mu^{q-1} + \ldots + \overline{B_q} = 0$$

are $\overline{\mu_1}, \overline{\mu_2}, \ldots, \overline{\mu_q}$. From (7.70) we have

$$Im\overline{\mu_k} < 0, \quad k = 1, \ldots, q.$$

By Corollary 7.1, the system (7.73), (7.77) has exactly $2(p+q)q$ linearly independent solutions (over the field of real numbers). Hence the general solution of this system is defined by

$$v(x,y) = \sum_{k=1}^{n} c_k v_k(x,y), \quad n = 2(p+q)q, \tag{7.78}$$

where $v_1(x,y), \ldots, v_n(x,y)$ are linearly independent solutions (over the field of real numbers) of the system (7.73), (7.77) and $c_1, \ldots, c_n$ are arbitrary real constants.

Now let us consider the equation

$$L_q v = 0, \quad (x,y) \in D, \tag{7.79}$$

where $v(x,y)$ is a real-valued solution to be found.

Passing in (7.79) to the complex conjugates and using the equality $\bar{v} = v$ we get $L^* v = 0$. Hence any real-valued solution of equation (7.79) is a solution of the system (7.73), (7.77).

According to Theorem 7.1, equation (7.79) has $m = q^2$ linearly independent real-valued solutions: $w_1(x,y), \ldots, w_m(x,y)$. Without loss of generality we can assume that

$$v_j(x,y) = i w_j(x,y), \quad j = 1, \ldots, m. \tag{7.80}$$

Substituting $v(x,y)$ from (7.78) into (7.74) and using (7.80) we get

$$u(x,y) = \sum_{k=m+1}^{n} c_k \omega_k(x,y), \quad (x,y) \in D, \tag{7.81}$$

where

$$\omega_k(x,y) = Re v_k(x,y), \quad k = m+1, \ldots, n. \tag{7.82}$$

We are going to show that the functions $\omega_k(x,y)$ $(k = m+1, \ldots, n)$ are linearly independent.

Let

$$\sum_{k=m+1}^{n} c_k \omega_k(x,y) \equiv 0, \quad (x,y) \in D, \tag{7.83}$$

where c_k $(k = m+1, \ldots, n)$ are some real constants.

From (7.82) and (7.83) we have

$$\sum_{k=m+1}^{n} c_k v_k(x,y) = iw(x,y), \tag{7.84}$$

where $w(x,y)$ is a real-valued function. Since $v_k(x,y)$ $(k = m+1, \ldots, n)$ satisfy equation (7.73), then $w(x,y)$ is a solution of this equation. Hence $L_q^* w = 0$, $L_q w = 0$ and

$$w(x,y) = \sum_{k=1}^{m} d_k w_k(x,y), \tag{7.85}$$

where d_k $(k = 1, \ldots, m)$ are some real constants. Substituting $w(x,y)$ from (7.85) into (7.84) and taking into account that the set of functions

$$\{v_k(x,y),\ iw_j(x,y): \quad j = 1, \ldots, m;\ k = m+1, \ldots, n\}$$

is linearly independent we obtain

$$c_k = 0, \quad d_j = 0, \quad (k = m+1, \ldots,\ n; j = 1, \ldots, m).$$

Therefore the functions $\omega_k(x,y)$ $(k = m+1, \ldots, n)$ are linearly independent.

Thus, any solution of the system (7.71), (7.72) is defined by (7.81) where c_k $(k = m+1, \ldots, n)$ are arbitrary real constants. Hence the assertion of Theorem 7.3 follows from (7.81).

Theorem 7.3 is proved.

Let Q_{pq} be the set of functions $v(x,y)$ admitting the representation

$$v(x,y) = L_q L_q^* u(x,y), \quad (x,y) \in D, \tag{7.86}$$

where $u(x,y)$ is any real-valued solution of equation (7.71).

One has the following

Theorem 7.4 *If the conditions (7.69), (7.70) are satisfied, then the dimension of the set Q_{pq} is equal to p^2.*

Proof. According to Theorem 7.1, equation (7.71) has exactly $(p+q)^2$ linearly independent real-valued solutions. The general solution of this equation in the class of real-valued functions is defined by

$$u(x,y) = \sum_{k=1}^{\rho} c_k u_k(x,y), \quad \rho = (p+q)^2, \tag{7.87}$$

where $u_1(x,y), \ldots, u_\rho(x,y)$ are the linearly independent real-valued solutions of equation (7.71) and $c_1, \ldots, c_\rho$ are arbitrary real constants.

Note that as $u_1(x,y), \ldots, u_\nu(x,y)$ we can take the linearly independent real-valued solutions of the system (7.71), (7.72). By Theorem 7.3 we have $\nu = 2pq + q^2$.

Substituting $u(x,y)$ from (7.87) into (7.86) we get

$$v(x,y) = \sum_{k=\nu+1}^{\rho} c_k v_k(x,y), \tag{7.88}$$

where

$$v_k(x,y) = L_q L_q^* u_k(x,y), \quad k = \nu+1, \ldots, \rho. \tag{7.89}$$

Let us show that the functions $v_k(x,y)$ $(k = \nu+1, \ldots, \rho)$ are linearly independent.

Let

$$\sum_{k=\nu+1}^{\rho} c_k v_k(x,y) \equiv 0, \tag{7.90}$$

where c_k $(k = \nu+1, \ldots, \rho)$ are some real constants.

Substituting $v_k(x,y)$ from (7.89) into (7.90) we get

$$L_q L_q^* u \equiv 0, \tag{7.91}$$

where

$$u(x,y) = \sum_{k=\nu+1}^{\rho} c_k u_k(x,y). \tag{7.92}$$

Hence the function $u(x,y)$ defined by (7.92) is a real-valued solution of the system (7.71), (7.72). We have

$$\sum_{k=\nu+1}^{\rho} c_k u_k(x,y) = \sum_{k=1}^{\nu} d_k u_k(x,y), \tag{7.93}$$

where d_k $(k = 1, \ldots, \nu)$ are some real constants.

Since $u_k(x,y)$ $(k = 1, \ldots, \rho)$ are real-valued linearly independent functions, it follows from (7.93) that $c_k = 0$ $(k = \nu+1, \ldots, \rho)$ and $d_k = 0$ $(k = 1, \ldots, \nu)$.

Thus, the functions $v_k(x,y)$ $(k = \nu+1, \ldots, \rho)$ are linearly independent.

From (7.88) it follows that the dimension of the space Q_{pq} is equal to p^2.

Theorem 7.4 is proved.

Remark. The results obtained in this chapter remain valid for multi-connected domains.

Bibliography

[1] L. Hörmander, *Linear partial differential operators* (Springer, Berlin – Heidelberg – New York), 1969.

[2] S. Agmon, A. Douglis and L. Nirenberg, *Estimates near the boundary for solution of elliptic partial differential equations satisfying general boundary value conditions, I. Comm. Pure & Appl. Math.*, **12** (1959) pp. 623–727.

[3] G. E. Shilov, *Mathematical Analysis, Second Special Course* (Moscow, 1965), (Russian).

[4] N.E. Tovmasyan, *Boundary value problems for partial differential equations and applications in electrodynamics* (World Scientific, Singapore, New Jersey, London, Hong-Kong, 1994).

[5] V.P. Palamodov, *Linear differential operators with constant coefficients* (Nauka, Moscow, 1967) (Russian).

[6] A.L. Pavlov, *On general boundary value problems for differential equations with constant coefficients in the half-space, Math. Sbornik* **103 (145)** (1977), pp. 367–391 (Russian).

[7] T.M. Kosheleva, *On a method of Cauchy problem resolution for partial differential equations in half-plane in the class of functions of polynomial growth, Izv. Vuz-ov, Mathematika* **1 (356)** (1992), pp. 25–32 (Russian).

[8] N.E. Tovmasyan, *General boundary value problem for a system of differential equations in half-space without of Lopatinsky condition, Differentsialnie Uravnenia* **20** (1984) pp. 132–141 (Russian).

[9] J.N. Vekua, *New methods for solving elliptic equations* (Noordhoff, Groningen, The Netherlands, 1953).

[10] G.E. Dikopolov, *On boundary value problems for differential equations with constant coefficients in half-space, Math. Sbornik* **50(101)** (1962) pp. 215–226 (Russian).

[11] N.E. Tovmasyan, *Boundary value problems for inexplicitly elliptic equations in half-plane, Pitman research notes in Mathematics series* **256** (1991), pp. 3–12.

[12] J.A. Bikchantaev, *Boundary value problem for elliptic equations with constant coefficients, Izvestia Vuz-ov, Mathematka* **6 (157)** (1975), pp. 56–61 (Russian).

[13] T.M. Kosheleva, *Non-correct boundary value problems for m-th order ordinary differential equations in the class of generalized functions, Izvestia Vuzov, Mathematka,* **3 (250)** (1983), pp. 51–58 (Russian).

[14] M.A. Lavrentiev and B.V. Shabat, *Methods of theory of complex variable functions* (Nauka, Moscow, 1970) (Russian).

[15] N.J. Muskhelishvili, *Singular integral equations* (Noordhoff, Groningen, The Netherlands, 1953).

[16] A.V. Bitsadze, *Boundary value problems for second order elliptic equations* (North-Holland, Amsterdam, 1968).

[17] I.N. Vekua, *Generalized analytic functions* (Pergamon Press, Oxford, 1962).

[18] V.V. Nikol'sky, *Electrodynamics and propagation of radio-waves* (Nauka, Moscow, 1978) (Russian).

[19] H. Buchholz, *Elektrische und magnetische potentialfelder* (Springer-Verlag, Berlin, Gottingen, Heidelberg, 1957).

[20] E.G. Kalashnikov, *Electricity* (Nauka, Moscow, 1977) (Russian).

[21] N.N. Mirolubov, M.V. Kostenko, M.L. Levinshtein and N.N. Tikhonov, *Methods of calculation of electric fields* ("Higher school" Publishing Co, Moscow, 1963) (Russian).

[22] Yu.Ya. Yossel, E.S. Kachanov and M.G. Strunsky, *Calculations of electric capacitances* (Nauka, Moscow, 1981) (Russian).

[23] V. Smight, *Electrostatics and Electrodynamics* ("Foreign Languages" Publishing Co, Moscow, 1954) (Russian).

[24] Y. Akira, *Calculations and measurements of electric capacitances, Dzenki Keysan* **38** (1970), pp. 156–161 (Japanese).

[25] A.N. Tikhonov and A.A. Samarsky, *Equations of mathematical physics* (Nauka, Moscow, 1953) (Russian).

[26] Kosaku Yosida, *Functional analysis* (Springer-Verlag, Berlin, Gottingen, Heidelberg, 1965).

[27] Krzysztof Maurin, *Metody Przestrzeni Hilberta* (Panstwowe wydownictwo naukowe, Warsaw, 1959).

[28] N.E. Tovmasyan, *Dirichlet-type problem for a class of high-order improperly elliptic equations, Izvestia AN Arm. SSR, Mathematika* **27** (1992), pp. 1–12 (Russian).

[29] G.S. Litvinchuk, *Boundary value problems and singular integral equations with the shift* (Nauka, Moscow, 1977) (Russian).

[30] A.I. Maltzev, *Foundations of linear algebra* (Gostekhizdat, Moscow, 1956) (Russian).

Index